EAST KENT
PLACE NAMES

The Homes of Kentish Men and Maids

Anthony Poulton-Smith

EAST KENT
PLACE NAMES

The Homes of Kentish Men and Maids

This edition published in Great Britain in 2013 by DB Publishing, an imprint of JMD Media.

ISBN 9781780912301

Printed and bound by Copytech (UK) Limited, Peterborough.

CONTENTS

INTRODUCTION

For years the history of England was based on the Roman occupation. In recent years we have come to realise the influence of the Empire did not completely rewrite British history, indeed there was already a thriving culture in England well before the birth of Christ. When the Romans left our shores in the fifth century the arrival of the Anglo-Saxons was thought to herald a time of turmoil, yet they brought the culture and language which forms the basis of modern England. Later the arrival of the Norsemen saw their influence and the same is true of our place names, the vast majority of settlement names in Kent are derived from the Saxon/Old English or Old Scandinavian tongues. There are also the topographical features such as rivers and hills which still have names given to them by the Celts of the pre-Roman era.

Ostensibly place names are simply descriptions of the location, or of the uses and the people who lived there. In the pages that follow an examination of the origins and meanings of the names in Kent will reveal all. Not only will we see Saxon and Scandinavian settlements, but Celtic rivers and hills, Roman roads and even Norman French landlords who have all contributed to the evolution to some degree to the names we are otherwise so familiar with.

Not only are the basic names discussed but also districts, hills, streams, fields, roads, lanes, streets and public houses. Road and street names are normally of more recent derivation, named after those who played a significant role in the development of a town or revealing what existed in the village before the developers moved in. The benefactors who provided housing and employment in the eighteenth and nineteenth centuries are often forgotten, yet their names live on in the name found on the sign at the end of the street and often have a story to tell. Pub names are almost a language of their own. Again they are not named arbitrarily but are based on the history of the place and can open a new window on the history of our towns and villages.

Defining place names of all varieties can give an insight into history, which would otherwise be ignored or even lost. In the ensuing pages we shall examine

2,000 plus years of East Kent history. While driving around this area the author was delighted by the unique place names found in this part of the county and so, having already taken a look at, among others, *East Sussex Place Names* and *West Sussex Place Names*, turned here to Kent in the form of *East Kent Place Names* and *West Kent Place Names*. This book is the result of the author's long interest in place names which has developed over many years and is the latest in a series which continues to intrigue and surprise.

To all who helped in my research, from the librarians who produced the written word to those who pointed a lost traveller in the right direction, a big thank you.

CHAPTER ONE – A

Acol

A name recorded in 1270 as Acholt. This name comes from Old English *ac holt* and describes 'the wood of oaks'.

Adisham

Found in 616 as Adesham and in Domesday as Edesham, this represents a Saxon or Jute personal name with Old English *ham* and describes 'the homestead of a man called Ead or Aeddi'.

Berwick is a common name principally because it does not have a specific meaning. From Old English *bere wic*, the first element is officially defined as 'barley', however it is also used as the generic name for grain crops, much as 'hoover' is heard more often than 'vacuum cleaner' today. Conversely the *wic*, correctly defined as 'specialised farm', is almost always a dairy farm. Here the two most common definitions, when put together, simply do not work and thus can only be said to be 'a barley farm'.

Blooden is found as Bledding and also Bledynge in 1270, this being 'the place of the family or followers of a man called Bledda'. First recorded as Bosingtune in 873, the Old English name of Bossington has a personal name with *ing tun* and describes 'the farmstead associated with a man called Bosa'. Cooting Farm is recorded as Cottinges in 1190, from Old English *cotingas* this is 'the place of the cottage dwellers'.

Pitt Wood, found as Petwode in 1332, comes from Old English *pytt wudu* and describing 'the wood where pits and hollows are found'.

Aldington

In Domesday this appears as Aldintone. Here the Saxon or Jute personal name is suffixed by Old English *ing tun* and tells of 'the farmstead associated with a man called Ealda'.

Locally we find Calley Well, a reminder of the family of John Coly who were recorded here in 1348. Copperhurst is from *copp hyrst* 'the top wooded hill' and is first seen as Cophurst in a document dated 1254. Poulton is a common place name and one always from *pol tun* or 'the farmstead by a pool'. Handen is a shortened form of the original *aet thaem hean denne* or 'at the high woodland', this has been shortened long before the earliest surviving record as Hendenne in 1334.

Ruffin's Hill takes the name of former residents Robert Ruffin and Henry Ruffyn, both recorded here in the same document of 1278. Named from the Shelley family, lords of the manor from the fifteenth century, to the south-west we find Shalley Wood. Stonestreet Green is named for a grassy area near a paved Roman road or *straet*.

The Good Intent public house recalls a schooner of this name which was known to be smuggling goods in and out of Kent during the 1830s.

Alkham

The earliest record of this name appears in a document from the end of the eleventh century as Ealhham. This features Old English *ealh ham* and descibes 'the homestead in a sheltered place', a name which perhaps should be understood as 'the sanctuary homestead'.

Chilton is a fairly common Old English place name, appearing at least three times in Kent alone. This is derived from *cilda tun* 'the farmstead of the children', although the understanding of 'children' is not what it is today for it applied to younger sons no matter what their age. This is particularly common in and around the south-east of England because of the Kentish law of Gavelkind, whereupon the land was inherited by all members of the family and not simply the eldest son.

Hougham is recorded in the Domesday book, this from a Saxon personal name and Old English *ham* and telling of 'the homestead of a man called Huhha'. Wolverton is seen as Wulfincton in 1226, here is a Saxon personal name with Old English *ing tun* and telling of 'the farmstead associated with a man called Wulfhere'.

Allington

Two places of this name in the county. This example, near Lenham, appears as Alnoitone in 1086 and as Eilnothinton in 1242. Here a Saxon or Jute personal name is followed by Old English *ing tun* and tells of 'the farmstead associated with a man called Aethelnoth'.

Locally we find the name of Midley, found in Domesday as Midelea and from *middle eg* and describing 'the middle island'. Note this is not an island in the modern sense but drier land in the marsh.

Appledore

The earliest records of this name date from the tenth century as Apuldre and as Apeldres in Domesday. This is a fairly common place name from Old English *apuldor* and tells us this was the '(place at) the apple trees'.

The local name of Park Farm is a remnant of the 'enclosed area for raising hunting animals' of Middle English *parke* here by the sixteenth century. Reading Street has two elements, together they tell us not only what to find here but whom. Listed as Reedinges in 1216, the first element is a tribal name meaning 'those who dwell in the clearing'. The addition, although not documented until the fourteenth century, tells it stands on a *straet* or Roman road. Rhee Wall is another example of Middle English where the final letter of the first element has migrated to the front of the second when it begins with a vowel. Here from Middle English *atter ee* meaning 'at the river' is a word followed by a reference to the dyke, a Roman construction to hold back the water in the marshland to the south.

Pub names of Appledore include the Black Lion which, as with almost all coloured animal names, is heraldic. Unusually we are fairly certain this is a reference to the consort of Edward III, Queen Philippa. She produced fourteen legitimate children with the king, including the highly popular Edward, the Black Prince.

Ash

Listed as Aesce at the end of the eleventh century, this name comes from Old English *aesc* and describes the '(place at) the ash trees'. Nearby New Ash Green is, as the name suggests, a new development and has borrowed the original name in the area.

Cop Street appears as Cobbistrete in 1366 and as Cobbestrete in 1487, both showing this was once associated with William Cobbe, recorded here with his family in 1348. No records of Durlock survive from earlier than the seventeenth century and yet, if this is from Old English *deor loca* or 'the enclosure for animals' as it seems, this is much older. Goss Hall is derived from the Old English *gos halh*, listed as Gosehal in 1202, speaking of 'the nook of land where geese are reared'.

Idleigh Court is found in the Domesday record of 1086 as Didele. Here Old English *leah* follows a personal name describing 'the woodland clearing of a man called Geda'. Knell Farm derives its name from Old English *elmas* which, like the main place name, refers to trees in 'the place of the elm trees'.

Molland Farm comes from Old English *mal land* and literally describes 'bargaining land'. This is understood as land where a rent was paid instead of a portion of the produce, effectively paying a tax in coin of the realm rather than a tithe. The Mote retains the original Middle English spelling of 'moat' but still refers to a defensive ditch. Overland Farm is named for its position at the *ofer land*, Old English for 'the cultivated land bordered by a bank'. Paramour Street is found from 1526, a reminder of the Paramor family recorded here since the fourteenth century.

Great Pedding Farm is first seen in a document dated 1251 as Pedinge, this from a Saxon personal name and Old English *ingas* and describing '(place) associated with the people of Pydda'. While no early forms of Redsteadle Wood exist, it is an interesting name as it seems to contain the Kentish word *steddle*, a dialect term which would then give 'the red framework of a timbered building'. From Old English *hreod leah* comes modern Ridley, the name describing 'the woodland clearing where reeds grow'.

Uphousden Farm is seen for the first time in 1484, recorded as Ophosen. This is from Old English *aet thaem uppe husum* and originally 'at the high houses'. Weddington appears as Wedinton in 1254 and as Wadyntone in 1348, this beginning as 'the farmstead associated with a man called Waedda'. Westfield Wood is a simple Saxon place name, originally described as 'the western open land' from Old English *west feld*.

Pub names begin with the Chequer Inn, which also gave a name to Chequer Lane on which it stands. It is one of the oldest known pub names in the world, an example found in the ruins of Pompeii beneath, rather ironically considering this place name of Ash, the ash of the eruption which destroyed this Roman city in ad79. Originally it showed a game similar to draughts was played on a chequerboard. Later it was also used to show the landlord was also a moneyer and the term is still used to refer to finance, for the man charged with holding the country's purse strings now holds the title of Chancellor of the Exchequer.

Ashford

Domesday's listing of Essetesford in 1086 points to this being from Old English *aescet ford* and speaks of 'the ford by a clump of ash trees'.

Bensted is recorded in Domesday as Benedestede, this is from *beonet stede* telling of 'the enclosed place where bent grass grows'. Clearly grass does not grow bent but will collapse under its own weight when it grows too tall, particularly in heavy rain or strong winds. This does enable us to deduce this region was enclosed but not for grazing, indeed we can assume nothing happened here if the grass grew so tall as to collapse. Unfortunately, as with many historical questions, finding the answers simply results in many more questions.

Ball's Wood and Ballsdown show a common origin, the name first recorded in 1636 as Bales Wood, the sole record of what was probably a local landowner. Bevenden is found as Benindene in 1278 and as Benyndenn in 1348, telling us this was 'the woodland pasture of a man called Bynna'. Godington has been found since the thirteenth century, a name describing 'the *tun* or farmstead of a man called Godda'. Cobb's Wood recalls the early seventeenth century when John Cobb was living at Ashford, the family can be traced back a further two centuries to Aldington. Foresham was brought here by a native the family of Stephen de Forshame, he is found in the parish records of Sutton Valence for 1351. Henwood is found as Henwude in 1272 and as Heanewode in 1313, showing this to be from Old English *aet thaem hean wudu* telling this was the '(place) at the high wood'. While early records of Wall Farm are not seen until the seventeenth century, the name is likely much older and from Old English *weall* 'a place by a wall'.

Podberry Wood is recorded as Pottebiri in 1294 and as Pottebery in 1305, showing this is from Old English *pott burh* or 'the fortified place where pots were once made'. While this defines the name the question remains as to the pottery, for the idea of a fortified place known for its pottery is not what we would expect. There have been countless numbers of pottery fragments found here, and that may well have been what the settlers here discovered when they moved in. If this is the case it shows this region was settled and then abandoned for a time before being resettled. Holm Wood is from Middle English *holm wude*, recorded as Holmwode in 1351 and referring to 'the wood of holly'. The name of Pearmain Way can be traced to a record of 1380 as Peremannesmed which speaks of 'the meadow associated with the family of a man called John Periman'.

The Locomotive Inn displays an image of George Stephenson's famous engine Rocket. The name refers to its location near the railway network and offers refreshment to travellers, perhaps even those alighting at Ashford International, the first stop for Eurostar trains arriving from Europe. Clearly the aim, having attracted your customer, was to sell the product. Many signs simply push the ale, such as in the case of the Hop Pickers. Another advertising ploy was to show you ran a patriotic establishment, hence the name of the Queens Arms or the male equivalent.

While the Golden Ball has been suggested as referring to Constantine the Great and thus early Christianity, in most cases it would allow landlords to produce a bold and visually pleasing yet simple sign. As today, landlords were keen to cut costs and painting a golden yellow circle as a sign could be achieved by almost anyone. In comparison the Bonny Cravat is a name which has many explanations, although few make any sense. If there is one explanation which stands out from the others it is the song which begins "*Jenny come tie my bonny cravat*", although any connection with this pub is unclear.

Aylesham

Although there are no records of this name before 1367, when it appears as Elisham, the place was certainly here before this. Again we find a Saxon or Jute personal name, this time with Old English *ham* and referring to 'the homestead of a man called Aegel.

Ackholt Road comes from a minor place name from *ac holt* 'the thicket by the oak trees'. Ileden is found as Gilding in 873 and as Gildinge in 1038, this being the '(place of) the family or followers of a man called Gilda'. Records of Ovenden are found from the fourteenth century although the name meaning 'the woodland pasture of a man called Hofa' is certainly very much older. Ratling describes 'the rubbish slope' from Old English *ryt hlinc*. Recorded as Rytlinge and Retlinga in the twelfth century, this is understood to describe 'rubbish' in the sense of 'rough growth or underbrush'.

CHAPTER TWO – B

Badlesmere

This name is found as Badelesmere in 1086, the Domesday record showing this is from Old English *mere* with a Saxon or Jute personal name. Here we find a name telling of 'the pool of a man called Baeddel'.

Woods Court took the name of the Wood family, who were lords of this manor from 1382. However this has only been known by this name since the seventeenth century. Although descendants of the Wood family were still living here by this time, prior to this the place was known by a completely different name. Since at least the thirteenth century the hamlet was known as Godislands, this from an earlier individual and a reference to 'the agricultural land of a man called Godd'. The minor name of Dryland Farm shows this was an area raised above the surrounding land, hence it drained more readily and was effectively 'drier cultivated land'.

Bapchild

The earliest surviving record of this name is as Baccancelde at the end of the seventh century. Here the Saxon or Jute personal name is followed by Old English *celde* and speaks of 'the spring of a man called Bacca'.

Barfreston

Domesday records this name as Berfrestone in 1086. Here the Saxon or Jute personal name is followed by Old English *tun* and tells us it is 'the farmstead of a man called Beornfrith'.

Barham

Early records of this name include Bioraham in 799 and as Berham in Domesday. Here the Saxon or Jute personal name precedes Old English *ham* and refers to 'the homestead of a man called Beora'.

Broome is a minor place name most noticeable as that of the golf course. Many years before golf had even been heard of in Kent it is found as Brome, this thirteenth century record from *brom* 'the place where broom grows'. Walderchain Wood can be traced to the thirteenth century with records such as Waterchine. This comes from Old English *weald wara cine* and refers to 'the cavern of the dwellers in the now cleared forest'.

Clip Gate Wood is recorded as Clipgate in 1662, a name which describes exactly what it seems 'the gate with a clip'. Names were coined to describe the place, showing something unique or at least made it stand out in the region. Today a gate without some means of fastening would stand out but five centuries ago it was a rarity to find a gate in the modern sense, the original *gaet* referred to the entrance or the way, not a barrier to provide or prevent access. Hence to find a movable portion of fencing (which is essentially all a gate is) was quite rare and to find it with a fastening or clasp would have been a talking point.

Derringstone is listed as Deringeston in 1262 and Deringgeston in 1321, a reminder this was 'the *tun* or farmstead of a man called Deoringe'. Upper Digges Place and Lower Digges Place were associated with the family of John Digges by 1254. Out Elmstead features a name meaning 'the *tun* or farmstead of a man called Uhhelm' with the later addition of 'out' informing of its outlying location. Ropersole Farm is seen as Roxpoll in 1444, this from *hroces pol* or 'the pool frequented by rooks'. It should be noted there is a chance the first element may also be used as the personal name Rook.

The Old Coach House is a pub name revealing this was a stop on the network which served the country for two centuries from the middle of the seventeenth century, the business killed off by the arrival of the railways. Adding 'old' is used in pub names to mean 'traditional', for clearly all coaching inns are old.

Bekesbourne

Listed in the Domesday survey as simply Burnes, the first form similar to the modern name appears in 1280 as Bekesborne. Here is a name from Old English *burna* or 'stream', today that being known as the Little Stour, to which we find the addition of lords of the manor from the twelfth century. Thus this name can be defined as 'the estate of the de Beche family on the river called Burna'.

Oakleigh Farm is a minor name from *ac leah* 'the woodland clearing among the oak trees', which dates back to at least 774.

Benenden

The listings of Bingdene in 994 and as Benindene in 1086 show this comes from a Saxon or Jute personal name and Old English *denn* and describes 'the woodland pasture of a man called Bionna'.

Backtilt Wood stands on a ridge of land between two streams, its name from *baec tilthe* 'the ridge of land which is tilled'. Cattsford is recorded as Crisford in 1313, Kirsforde in 1327, and Kersford in 1334, a name from Old English *cresse ford* 'the river crossing where cress grows'. Chittenden describes 'the woodland pasture of a man called Citta', the Saxon or Jute personal name followed by Old English *denn*.

Crit Hall is found as Crotehole in 1292, Crutehole in 1402, and Crittall in 1690. Here the personal name is suffixed by *hol* and describes 'the hollow of a man called Crotta'. Dingleden appears as Tenelyngden in 1226 and Thenglesdenn in 1383, these records showing this 'the woodland pasture of the prince', a Saxon or Jute royal estate. Seen since the thirteenth century, Dockenden describes 'the woodland pasture of a man called Dokel'.

Frog's Hole appears as Ffroggenhole in a document from 1262, the name from *froggena hol* or 'the hollow frequented by frogs'. Matilda Godard was at Goddard's Green in 1348. Hinksden is derived from *hengest denn* meaning 'the woodland pasture where stallions are grazed'. Hole Park is recorded as Hole in 1278, this from *hol* meaning 'hollow'. Walkhurst Road can be found in the thirteenth century as Walcherst and Walkeherst. This comes from Old

English *hyrst* and a Saxon personal name giving 'the wooded hill of a man called Wealca'.

Iden Green comes from Old English *iw denn* and tells of 'the woodland pasture marked by yew trees'. Ramsden appears as Ramesden in the twelfth century, from *rammes denn* this is 'the woodland pasture where rams are reared'. Standen is recorded as Stanehtandenn in 858, this from *aet thaem stanihtan denne* telling us it was 'at the stony woodland pasture'.

Tilden is derived from Old English *telga denn* or 'the woodland pasture where young trees grow' and likely evidence of a planned plantation. In Tottenden Wood we see the same suffix, here *denn* combines with a Saxon personal name and speaks of 'the woodland pasture of a man called Tota'. With Willerd's Hill Wood we can find the name coming from the former residents in 1348, when the family of Thomas Willerd were known to be here.

Bethersden

The earliest surviving record is as Baedericesdaenne in a document from the end of the eleventh century. This features a Saxon or Jute personal name and Old English *denn* and refers to 'the woodland pasture of a man called Beaduric'.

Brickhurst Wood was home to a former kiln for making bricks, the *hyrst* telling of 'the wooded hill'. Butcher Wood comes from the name of the family of Walter Butcher, here by 1340. From Old English *haecces denn* comes Etchden Farm or 'the woodland pasture reached via a hatch gate'. Lovelace Farm can be traced back to the middle of the fourteenth century and the family of John Lovelace.

Runsell Farm is found as Rynsole in 1254 and Rynsoll in 1527. These show the origin to be *ryne sol* or 'the muddy pool near running water'. Old Surrenden Manor Lane features the element *denn* with a Saxon personal name and refers to 'the woodland pasture associated with a man called Swithred'. Nearby Surrenden Dering has taken the basic place name and added that of the family of Richard Dering, who was here by 1480, and which also explains the need for the addition of 'Old'. In 1644 one Clara Toke is recorded at what is now known as Tokeland Wood. With records of Wychyndenn in 1332, Wissenden speaks of itself as 'the *denn* or woodland pasture of a man called Wicca'.

Here we find the Bull Inn, most often heraldic or simply a favourite animal, but here it comes from Bull Hill, itself named from where such were grazed. The Royal Standard refers to the flag representing the monarch, and thus a royalist establishment.

Betteshanger

Found as Betleshangre in 1176, this name comes from Old English *bytle hangra* and speaks of 'the wooded slope by a building'.

Bicknor

Not seen before the record of Bikenora in 1186, this name comes from Old English *bica ora* and tells of 'the slope below the pointed hill'.

Here we find Gorham Wood, a name describing 'the homestead at the wedge-shaped area of land' and derived from Old English *gara ham*. Recorded as Garham in 1185, it was not until much later the name was used for the woodland.

Biddenden

Here the Saxon or Jute personal name precedes Old English *ing denn* and refers to 'the woodland pasture associated with a man called Bida'. The name is recorded as Bidingden in 993.

Ayleswade is first seen in a document from 1224, where it appears as Halsnod. This represents Old English *halh snad* 'the corner of the detached woodland'. Bettenham describes 'the hemmed-in land of a man called Betta', a shortened version of the name was transferred to Bettmans Wood Farm.

Fosten Green comes from *forst denn* 'the frosty woodland pasture', although 'frosty' is probably used to mean 'unwelcoming' as sub-zero temperatures are not frequent enough.

Found as Berkelegh in 1327, Birchley comes from *beorc leah* or 'the woodland clearing in or by the birch trees'. Listed as Bicoppesdenn in 1226 and as Bisshopisdenn in 1313, Bishopsdale refers to 'the woodland pasture associated with the bishop'. Castweazel is recorded as Cassetuisle, Castwysle, and Castwysele between the twelfth and fourteenth centuries. Here is a name from Old English *casan twisla* or the '(place) of the hovel dweller in the fork of the streams'. Lashenden puts Old English *denn* after a personal name to tell of 'the woodland pasture of a man called Laecca'.

Poors can be traced back to the fourteenth century and a family sometimes given as Poor and alternatively Por.

Rogers Wood recalls the family of Robert Roger, here in Biddenden in 1357. Nearby Rogley is derived from *ruh leah* meaning 'the rough woodland clearing'. Tuesnoad is first seen in the thirteenth century, this a Saxon personal name and Old English *snad* speaking of 'the detached area of land of a man called Tiw'. A name meaning 'the woodland pasture of a man called Waecca', Washenden is recorded as Wecchynden in 1270 and as Wachingdenn in 1278. Woolpack can only have come from a former inn, itself showing a link to the wool trade. Worsenden has been seen since the thirteenth century, the name telling of 'the *denn* or woodland pasture of a man called Weorca'.

At Three Chimneys the Three Chimneys public house only has two for the place name is a humorous reference to the fingerpost

Bilsington

Domesday records this name as Bilsvitone in 1086. Here a Saxon or Jute personal name precedes Old English *tun* and speaks of 'the farmstead of a woman called Bilswith'.

Locally we find Poundhurst, first seen in 1240 as Pundhurste and from Old English *pund hyrst* and telling of 'the pound for animals by the wooded hill'. Swanton comes from *swana tun* and tells of 'the farmstead of the peasants'. Wallsfoot Sewer began as Wallesende, as the name says it could be found 'at the end of the wall'. Later the drain or sewer was dug along the foot of this wall, thus the present name speaks of 'the sewer or drain at the foot of the wall'.

Birchington

Listed as Birchenton in 1240, this comes from Old English *bircen tun* and describes 'the farmstead where birch trees grow'.

Brooks End speaks for itself, the name recorded as Brokesend in 1205 and as Brockesende in 1292. Down Barton Farm lies on a 'hill or down', the main part of the name describing 'the barley farmstead' from *bere tun*. Upper Hale, recorded as Halle in 1292, and Nether Hale, seen as Le Nedyr Hall in 1456, share an origin in Old English *heall* or 'manor house'. Potten Street, also seen in the oddly-named Potten Street Road, derives its name from the family named Potyn, here by the fourteenth century.

Quex Park is seen as Queax in 1677, the place derived from the family who held the land and represented by Stephen and Robert le Queke. Shuart is a name from Old English *scyrte* and speaking of 'the area of land cut off' (literally skirted).

Bishopsbourne

A name with similarities to Bekesbourne for it also refers to the Little Stour before it was known as such. Documented as Burnan in 799, Burnes in 1086, and Biscopesburne in the eleventh century, here the Old English *burna* or 'stream' is joined by *biscop* and tells us this was 'the estate of the Archbishop of Canterbury on the river called Burna'.

What is now known as Benhill Wood took its name from the place on which it stands, that being known as 'the slope of a man called Beonna'. Gorsley Wood is found as Gosley in 1541, these forms are rather late and it is difficult to see if this is *gorst leah* 'the woodland clearing where gorse grows' or *gos leah* 'the woodland clearing where geese feed'. Historical records of Great Bursted Farm give the name as Beristede, Burstede, and Borstede, all of these in the thirteenth century. This is derived from *burh stede*, 'the place at the stronghold'. From Old English *lang hamm*, Langhampark Lodge describes its location at the 'long hemmed in place'.

The Mermaid public house is probably heraldic, an indication the family had links to the sea.

Blean

While Domesday lists this name as Blehem in 1086, over three centuries earlier in 724 a document gives this name exactly as it appears today, making it at least one of the oldest unchanged names if not the oldest. This undoubtedly comes from Old English *blea* and describes the '(place in) the rough ground'.

Amery Court is derived from Old French *almosnerie* telling us this was the site of an 'alms house'. Both Clowes Farm and Clowes Wood share an origin in Old English *clus* describing 'the woodland enclosure'. Listed as Gremyngehelde in 1313, Grimshill Wood is from a Saxon personal name and Old English *inga hielde* to speak of 'the slope of the family or followers of a man called Grimm'. The name of Pean Hill is derived from *pund*, an Old English word meaning 'pound' or 'enclosure for animals'.

The Hare and Hounds public house can only show a link to hare coursing, this blood sport banned in 2005. To the dog owners the exercise was more about their animals turning and chasing the quarry than a kill; for the hare invariably escaped.

Bobbing

A document dating from the end of the eleventh century gives this name as Bobinge. Here a Saxon or Jute personal name and Old English *ingas* together tell of the '(place of) the family or followers of a man called Bobba'.

Bonnington

Domesday records this name as Bonintone, showing this is a Saxon or Jute personal name with Old English *ing tun* and telling of 'the farmstead associated with a man called Buna'.

Borden

Listed in 1177 as Bordena, this name comes from Old English *bor denu* and describes 'the woodland pasture by a hill'.

The suffix of the local name of Eyhorne Hatch is from Old English *haecc*, a 'hatch gate' being a panel across a way. The first part can be traced to Saxon times when this was the meeting place for the local hundred of Eyhorne, a name from *aet thaem haegthorn* telling of this being 'at the hawthorn tree'. Oad Street appears as Holdestrete in 1254, from *eald straet* and meaning 'the old Roman road'. Sutton Baron is another example of the common *suth tun* or 'southern farmstead' with the addition actually a corruption of the parish name, having been added for distinction.

The pub name began as an advertisement, a visual image carrying a message which could not be read as the vast majority were illiterate. One of the most common names in the country is named after the plough, a name inviting those who worked the land which, until the Industrial Revolution, was virtually everyone. Such a common name produces its own problems, if there is a Plough in every village it would be difficult to know which one was being referred to. Thus landlords looked for a unique addition and one of the first was another implement, the harrow – a plough turns the soil, the harrow breaks up the large clods into a finer tilth to make it easier for the crops to gain a foothold. Thus we find the Plough and Harrow at Borden.

Once a common sight on the village green, the Maypole is a reminder of the feasting and festivities of a traditional May Day, with the main attraction the dancing and the interweaving of coloured ribbons around the maypole.

Boughton Aluph

Records of this name include Bocton in 1020, Boltune in 1086, and Botun Alou in 1237. The basic name comes from Old English *boc tun* and refers to 'the farmstead held by charter'. Here the addition is manorial and points to a thirteenth century lord of the manor called Aluf. Nearby Boughton Lees, with the same basic origin, features the suffix *laes* meaning 'pasture'.

Buckwell Farm comes from *boc wielle* or 'the spring by the beech trees'. Goat Lees is recorded as Gatele in 1247, from *gat leah* this tells of 'the woodland clearing where goats are kept'.

Boughton Malherbe

As with the previous entry this name comes from Old English *boc tun* and refers to 'the farmstead held by charter'. This name is recorded as Boltune in 1086 and as Boctun Malerbe in 1275, here the addition is again manorial and refers to the Malherbe family who held the manor from the thirteenth century.

Bowley Farm is listed as Bogelei in Domesday, this telling of the *leah* or woodland clearing of a woman called Bucge'. Chilston Park comes from Old English *cildes tun* or 'the farmstead of the children', not that children meant what it does today, here they refer to their heirs and should probably be defined as 'young men'. Wallet Court comes from *wega gelaete*, quite literally 'the meeting of the ways' and an ancient junction recorded since at least the fourteenth century when it is found as Woghelete.

Coldbridge Wood and Coldbridge Farm share an origin from *aet thaem coligan brycge* describing its location 'at the blackened bridge'. While this is undoubtedly the origin, understanding its meaning is a different matter entirely. It may refer to the blackened appearance of a feature associated with or near the bridge, perhaps stones on the bed of the stream. Although there is a Blackpit Wood nearby, its name a reminder that charcoal was burned here, which may have influence the name of Coldbridge too.

Southpark Wood is self-explanatory, referring to parkland to the south of the parish. Thornden comes from *thorn denn* and recorded as Thornden as early as 850, the name describing 'the woodland pasture marked by thorn bushes'.

Boughton-under-Blean

A name recorded as Boltune in 1086, Bocton in 1226, Bocton juxta la Blen in 1288, Boctone juxta Bleen in 1292, as Boghton under Blee in 1386, and as Boctone under Bleane in 1610. An interesting name for the late fourteenth century record is found in Geoffrey Chaucer's *The Canterbury Tales*. As a place name the basic name, as the previous entry, is from *boc tun* 'the farmstead held by charter'. The addition describes its position beneath the Forest of Blean, described under its own entry. Nearby Boughton Street, now taken as a place name, is first seen as Bocton Street, the road leading to (or from) Boughton-under-Blean.

Both Cleve Hill and Cleve Marshes share a beginning from Old English *cliff* telling us these can be found 'at the cliff or bank'. Colkins is a manorial name, a reminder of John and Richard Colkyn who were here in the fourteenth century. Dane Court was built in 'a valley', and that is what the name means, coming from the Old English *denu*. Fairbrook Farm holds no surprises as 'the beautiful stream', although the Old English description of *faeger broc* is more likely to refer to the vegetation than to this small tributary itself. Hickmans Green derives its name from the family said to be 'tenants of the land' in a document dated 1708. Oversland shows this land was a possession of Clement Oueray by 1367. In 1278 the family of August Trente were living at what is now known as Trent Wood.

Pub names begin with the Queens Head which, as with the male version of Kings Head, often depicts a particular monarch. However the message is simply to show this establishment supported the monarchy, in effect they were patriots.

Brabourne

Early records of this name include Bradaburna in 860 and as Bradeburne in 1086. A name from Old English, where *brad burna* tells us it was the '(place at) the broad stream'. There is also an East Brabourne and a West Brabourne.

The local name of Goldingbank Wood is derived from the family of William Goldyng, recorded here in 1348. Hampton is one of the most common English place names, indeed if this was a larger place we would expect an additional element to make it unique. Here is Old English *ham tun*, understood as referring to 'home farm'. Monks Horton appears as simply Hortun in 1086, this from *horh tun* 'the dirty or muddy farmstead'. The addition is not recorded before 1610 and simply shows this as a possession of Horton Priory. Quarrington is found as Cudrinton in 1254, here the Saxon personal name precedes *tun* and speaks of 'the farmstead of a man called Cuthhere'.

St Peter's Church at Monks Horton

Bredgar

Found in a document from around the end of the eleventh century as Bradegare, this name comes from Old English *brad gara* and speaks of 'the broad triangular plot'.

Chantry Cottages was originally the chantry house of the College of the Holy Trinity it was founded by Robert de Bradgare in 1393. The 'group of buildings on a hill' is recorded as Dungesell in 1225 and today is known as Deans Hill.

The Sun Inn is an example of one of the most simple of pub names. Almost anyone could produce a sign with a yellow ball with a few rays coming from it, the message being one of a warm welcome.

Brenzett

Recorded in Domesday as Brensete, this name is derived from Old English *berned set* and tells us this was known as 'the burned fold'.

Court-at-Wick to the north describes 'the manorial *wic* or specialised farm', that speciality nearly always dairy produce.

Bridge

Even today it is easy to see this name as describing the '(place at) the bridge'. This name comes from Old English *brycg* and appears in Domesday as Brige.

Locally we find Renville, a name recorded as Trimfeld in 1240, Thremfelde in 1332, and as Renfelde in 1535. These show forms from the sixteenth century to be much corrupted and should have become Threnfield today. Hence this is 'the trimmed open land' from old English *thryme feld*.

Broadstairs

The earliest known record of this name dates from 1435 as Brodsteyr. This name comes from Old English *brad staeger* and speaks of 'the broad stairway or ascent'.

Locally we find Buddles, derived from the family of Henry and Robert Buthel who were here around 1275. There are plenty of records of Dumpton from the late twelfth century, all pointing to an origin of 'the farmstead associated with a man called Dudeman'. Sakkett's Hill features a surname, the families of John and Walter Zaket here in the middle of the thirteenth century.

The Charles Dickens public house honours one of our most-loved writers. Best remembered for his *A Christmas Carol* he wrote much more, including 20 novels, numerous short stories, a little poetry, some non-fiction, and even the odd play. He also found time to produce children; Charles jnr, Mary, Kate, Walter Landor, Francis, Alfred D'Orsay Tennyson, Sydney Smith Haldimand, Henry Fielding, Dora Annie, and Edward.

Captain Digby took its name from a man who led a campaign against smugglers along the coastline of Kent in the nineteenth century. A more famous mariner gave his name to the Lord Nelson, a man with more pubs in this country named after him than any other individual. Another maritime name is the Tartar Frigate, it being named after the eighteenth vessel of the Royal Navy given this name. A Tartar, someone from Tartary, was given to just about anyone seen as overly violent and most uncouth.

Named from the novel by Charles Dickens, the Barnaby Rudge seems a strange choice of name, for the eponymous character is not particularly inspirational, this pathetic and impractical individual seemingly more concerned with how the stars twinkled rather than the political intrigue of the Gordon anti-popery riots, the main theme of the book. The Richmal Crompton remembers another novelist, Richmal Crompton Lamburn is remembered for her *Just William* stories and associated with Bromley as she died here in 1969.

In the Crown and Thistle is a name commemorating the union between England, represented by its monarchy, with the thistle of Scotland. No surprise to find a golf course near the pub named the 19th Hole, this being the traditional name for the club house where golfers would gather for a drink after their game.

Brook

Another name which is easy to define from the modern form. Appearing in an eleventh century document as Broca, this name comes from Old English *broc* and refers to 'the brook'.

Much less obvious is the pub named the Honest Miller. When all flour was milled locally the miller had a ready, and ostensibly inexhaustible, market. His services were indispensable, unless a person had endless hours free to spend grinding corn by hand, and it is a sad fact that millers exploited this need and were renowned for their dishonesty. Indeed it is probably true that any miller felt duty-bound to syphon a little off for himself and his family. Thus in the Honest Miller we find a name which suggests something very special and possibly unique.

Burham

Found as Burhham in the tenth century and as Borham in Domesday, this is from Old English *burh ham* and tells us of 'the homestead near the fortified place'.

Pub names include the Fleur-de-Lys, this a flower seen in heraldry and representing France.

Burmarsh

A name which can be found in Domesday as Burwarmaresc and early as Burwaramers in the seventh century. Here Old English *burh ware mersc* speaks of 'the marsh of the town dwellers', that 'town' being the city of Canterbury. Abbott's Court was associated with St Augustine's Monastery at Canterbury.

The delightfully sounding name of Willop Sewer appears as Wylhope in 1253 and as the modern form as early as 1271. Willop comes from *willig hop* and tells of 'the enclosed marshland where willows grow', with the less attractive part simply describing the channel along which the marshland drains.

In the Shepherd and Crook we find a name which is clearly a reminder of the wool trade, although it also has religious associations.

CHAPTER THREE – C

Calcott

Found south of Herne Bay, this place is seen as Kaldecote in 1250 and Caldecote in 1357. Here is a common place name found across England and from *cald cot* meaning 'the cold or exposed cottages'.

Canterbury

An important settlement in England for centuries, it is recorded as Canwaraburg around 900 and as Canterburie in the Domesday record of 1086. This is a name which tells us of 'the stronghold or fortified town of the people of Kent', an ancient Celtic name combines with Old English *ware burh*.

Canterbury Castle

Local names begin with Dane John, a name from Old French *donjon* and meaning 'dungeon'. This is not a dungeon in the modern sense, this refers to a defensive mound with the city walls and later a public garden. The Friars is first recorded in 1690, this recalling a priory founded by a Carmelite Order in 1240. Hammonds Corner was associated with the family of William Hammond, first recorded here in 1270. Wytherling Court is recorded as Wetherlinga in 1174, this being the '(place) of the family or followers of a man called Wither'.

Many old towns and cities retain the names of the old gates, such as the example here of Northgate recorded since 1087. Another is Queningate, found in a document of 762 and clearly referring to 'the queen's gate'. Westgate Court is at the site of the western gate of the city, marking the site of the Westgate Hundred. Similarly Worthgate tells of 'the gate of the people of the Wye'. In Wincheap Street we find a description of the place. Here, as seen in Old English *waen ceap*, is 'the wain market' and literally where the waggons from the market are loaded. In Beer Cart Lane the name is of obvious derivation and later gave a name to the pub named the Beer Cart Arms. It is clear the pub was named afterwards as there are no 'arms' granted in this example.

Public houses include the Grove Ferry Inn, named after the hand-drawn ferry which came past the pub when it was first built in 1831. Merton Farm occupies land which was once named from Old English *mere tun* and speaks of 'the farmstead by the pool'. Scotland Hills has nothing to do with lands north of the border, this is *scot land* or 'land on which a tax or payment was due'. The phrase to get off 'scot free', ie without having paid full dues, is the only remaining use of the word. With Yockletts Farm comes a name from Old English *geoclad* and speaks of a measurement of land area, this recorded as Yoklete in 1226.

Outside the Bell and Crown public house is a sign showing this pub shows allegiance to the two most powerful and largest landholders in the country, the church and the monarchy respectively. In the Maiden's Head we find a pub name found in several places in Kent, although not exclusively to the county. All were likely suggested by the town of Maidenhead and, although the meaning of 'the stone of the maidens' has never been understood, is not restricted to that place but found as a minor name elsewhere. Whether this represents an old minor name or not is not clear but, if it does, it makes the question of understand its meaning even more intriguing.

In the Hop Pocket we have a pub name referring to one of the main ingredients in brewing beer, along with the standard volume of one and a half hundredweights contained in the sack known as a 'pocket'. In the Tap and Spile we have a name referring to the method used to draw ales from a barrel. While the 'tap' is obvious the 'spile' requires some explanation. This is simply a wooden peg, one which plugs a hole at the top of the barrel. It is removed before the tap is opened and replaced after it is closed to enable air to replace the volume of liquid removed and thus give a smooth and even flow from the tap.

The Leopards Head is a reference to the Weavers' Company who use this in their coat of arms. The Old Weavers' House is a more obvious reference to the same workers. The Thomas Ingoldsby pub was named for the *nom de plume* of Richard Harris Barham, who was born in Canterbury in 1788.

The Thomas Becket remembers the man, murdered by knights in Canterbury Cathedral, whose tomb became a place of pilgrimage. A reference to the *Canterbury Tales* was used to produce the name of the Franklin and Firkin. Once the Firkin Brewery's pubs would always use their name as the second part of the pairing, with the first an example of alliteration relevant to the area, although sometimes the initial 'F' became a 'Ph' when necessary. In Canterbury the two are linked by Geoffrey Chaucer's book, for Franklin was one of the characters making the pilgrimage. He being the Squire and the subject of the sixth tale. Another pilgrim gave a name to the Gentil Knyght, this pub using Chaucer's original spelling of 'gentle knight'.

What seems another religious name is the Bishop's Finger. Yet this pub is named from a slang term for what we would refer to as a fingerpost.

Capel le Ferne

A place name found in a number of counties, always from Middle English *capel* meaning 'chapel'. The name is recorded as Capel ate Verne in 1377, the addition meaning 'at the ferny place' from Old English *ferne*.

The minor name of Fobles Wood, a name first seen as Fobles Farm in 1407. While there is no record of a family of this name living here there can be no other explanation. Lilley Farm is found since the early thirteenth century, from *lind leah* this is 'the woodland clearing by the lime trees'. Ploggs Hall was associated with Richard Plog, the family recorded as here by 1240. Satmore is first recorded in 1275 as Shottemer, from Old English *sceota mere* and speaking of 'the shoat or trout pool'.

At the Valiant Sailor we find a variation on 'soldier', a name which came to prominence following the English Civil War.

Challock Lees

The earliest surviving record of this name dates from 824 as Cealfalocum. The modern name features three Old English elements, *cealf loca laes* speaking of 'the enclosure of the pasture for calves'.

Bodshead Farm has no surviving records prior to 1778 when it appears as Badsted. These late forms mean we have to guess at the evolution, this probably beginning as 'the settlement of a man called Baeddi' although should earlier records be found in the future this may be confirmed or amended. High Snoad Wood is not recorded until 1535, however the name appears to be much older from *heah snad* and, very specifically, referring to 'the high area of detached woodland'. Molash can be traced to Old English *mael aesc*, literally 'speech ash' and referring to an ash tree which marked the location of a meeting place.

Charing

Listed as Ciorrincg in 799 and as Cheringes in the Domesday record of 1086, this name probably comes from Old English *cerring* meaning 'a bend in the road'. However there is a possibility this may be two elements, a Saxon or Jute personal name and Old English *ing* and speaking of the '(place) associated with a man called Ceorra'.

Acton Farm comes from 'the *tun* or farmstead of a man called Acca'. Brockton is found as Brokton in 1286, this from *broc tun* 'the farmstead near marshy ground'. Burnt Barn is first recorded as such in 1799, a fairly good indication the fire occurred shortly before this. As a place name Hawk's Nest is very recent, indeed as the name appears for the first time in 1782 it probably has no etymological value but was simply created.

With records from the ninth century, Housendane Wood speaks of 'the *denn* or woodland pasture of a man called Husa'. Impkin comes from Old English *ing mere* with a personal name and refers to 'the pool associated with a man called Aemic'. Newcourt Wood is a combination of Middle English *newe* and French *court* to speak of 'the new mansion'. While the name for a large house may seem obvious, it is actually the 'new' which is more important as it shows there was an earlier building here, one which was known by another name and which has yet to be discovered.

Newland is self-explanatory, although from Old English *niwe land* 'the land newly taken for agriculture' has not been 'new' since at least 1264. Ray Wood and nearby Raywood Farm derive their name from Middle English *atter eye* 'at the island'. Here the last letter has migrated from the first word to the second, and when the element *atter* to be followed by a vowel this is not only commonplace but to be expected. Rigshill is found as Rikesole in 1278 and Riggesole in 1367, these from Old English *sol* and a Saxon personal name telling of 'the muddy pool of a man called Rica'.

At the Bowl public house the reference is to a punchbowl. This remained the property of the village and was filled and the contents distributed to residents when the occasion demanded. While the idea of punch may seem a twentieth century idea, it was first brought to Britain in the seventeenth century courtesy of those vessels carrying cargo on behalf of the East India Company. The word 'punch' comes from Hindi where *panntsch*, meaning 'five', was made from a spirit, sugar, lemon, water and tea or spices.

Prior to the building of the canals, heavy loads had to be transported by road. This meant large draught horses pulling the loads along little more than dirt tracks, made even more difficult as they were often badly rutted and extremely muddy. Both horses and drivers required regular rest stops, these being pubs and giving a name to the Wagon and Horses. This is an unusual example as almost all pubs have the alternate spelling 'waggon' which is still the preferred spelling, although one 'g' is an acceptable alternative.

Chart (Great & Little)

Two places with a common origin, the additions requiring no explanation. These places are recorded as Cert in 762, Certh and Litelcert in 1086, and Magna Chert during the thirteenth century. Time and again these records show the Old English element *cert*, a word meaning 'rough ground'.

Two signs for Great Chart, the former erected as part of the millennium celebrations, the other a traditional road sign

Boyce Wood can be traced back to a document from 1664 when the family of Edward Boyse were at Boyes Hall, itself named from its builder Thomas Boys. Bridewell Plantation may not be found before the early seventeenth century, yet the name is much older and from *bridd wielle* 'the spring frequented by young birds'.

Buxford House was built at 'the ford frequented by bucks', from Old English *bucca ford*. Calehill is a reminder of this being the meeting place for the Calehill Hundred, this being from *calu hyll*, Old English for 'the bare hill'. Chilmington Green is found as Chelminton in 1226, this describing 'the *tun* or farmstead of a man called Ceolhelm'. Daniel's Water was known as Daniells Water in 1634, deriving its name from the family of John Daniel who was here by 1346.

Eggringe Wood is a reminder of 'the family or followers of a man called Eadgar'. Goldwell comes from *golde wielle*, Old English for 'the marigold spring', one where such flowers would have grown in abundance. Newstreet Farm shows it lies on the 'new street' from Middle English *newe strete* and, unlike Old English *straet*, does not refer specifically to a Roman road. Ninn Lodge is a Middle English name, one which originally described its location *atten inne* or 'at the inn'. Pigsbrook Wood probably explains itself as 'a wetland where pigs were kept', yet it is also possible this was brought here as a surname for the family of Thomas Pykebroke were here by 1348. Either way the meaning is the same.

Rooting appears as Rotinge in the Domesday record of 1086, this telling us it was the '(place) associated with the people of Rota'. Thanet Wood is not named after the place but a former lord of the manor, Nicholas Tufton was created Earl of Thanet in 1629. Worten Farm appears as Werting in 1202, this speaking of the '(place) of the family or followers of a man called Wierta'.

In the Swan Inn we have a pub name derived from one of two equally likely sources. Either this is a direct reference to this majestic bird or this comes from a coat of arms, the image chosen for the same aesthetic reasons.

The Hoodener's Horse is a delightful name which owes its existence here to Kent's most famous crop, hops. Hooding the hops refers to the process of drying them in the oast house. This was also the time when morris dancers would use

the hooden horse as part of their performance, a prop similar to the hobby horse.

Hoodener's Horse public house in Great Chart

Chartham

The similarity between this and the previous entry is clear, and for very good reason for the Old English origin here is *cert ham* or 'the homestead in rough ground'. The name appears as Certham in 871 and as Certeham in 1086.

Denstead Farm and Denstead Lane take an early place name, from *denu stede* this is 'the place of the valley'. Howfield Farm can be traced to a Saxon personal name and Old English *feld* and telling of 'the open land of a man called Huhha'. From Old English *hund stede*, Hunstead Wood tells of 'the place where hounds are kept'. Describing 'the farmstead with a mill' Milton is a common enough name although one usually seen with a second element, which would undoubtedly be present if the place were bigger.

Mystole appears exactly as the modern form in the earliest record from 1610. This comes from Old English *meox steal* and describes 'the place of the dung stall' – a rather negative way of describing a cattle shed. Pickelden Farm is seen as Pykyndenne in 1327, a Saxon personal name suffixed by *denn* and telling us this was 'the woodland pasture of a man called Pic'. Records of Thruxted are not found before the sixteenth century, yet if the suspected origin of *thrysce stede* or 'the place where thrushes are seen' is correct then the name is much older.

Shalmsford Street, the final element refers to the nearby Roman road, takes its basic name form *scameles ford* or 'the ford by a shambles'. Defining the name describes a lot more of the area than is first apparent. The ford crosses the Great Stour and, with the remainder of the name referring to a slaughter house, the blood and other unwanted remains will have drained into the river and away to the sea.

Chilham

Records of this name include Cilleham in 1002 and 1086. These records do not allow this name to be defined with any certainty. Most likely this is a Saxon or Jute personal name with Old English *ham* and referring to 'the homestead of a man called Cilla or a woman called Cille'. Alternatively the first element may be

Old English *cille* which would give a definition of 'the homestead by a spring'.

Bagham is a local name which appears to be associated with John Bagge of Chilham, whose family were here by 1309. Barty Farm and Barty House share a name from Old English *burh teag* 'the enclosure of the stronghold'. Bowerland is found as Burlonde in 1275, coming from *bur land* 'the agricultural land by or of the dwelling place'.

Thornham is easy to see as 'the homestead of or near the thorn bushes'. Corbier Hall Wood is named after the family of Robert Corbie, lord of the manor of Thornham. East Stour Farm speaks for itself as being found 'east of the River Stour'. The Hundred of Felborough is still seen in the name of Felborough Wood, originally from *fealu beorg* or 'the mound near fallow land' this will have been the meeting point for the hundred. Mountain Street is not named from any precipitous incline but linked to the family of de Opmanton, in turn from a place name meaning 'up men's farmstead'. It remains to be seen whether the surname gave the modern place name or if both are derived from the earlier place name.

Pole Wood is seen as Polledewode in 1338, coming from Middle English *pollede wode* speaks of 'the wood where pollarding was seen'. Pollarding involves cutting the vegetation from the tree trunk above the height reached by browsing animals, this encourages the growth of new straight poles for use in building. Coppicing will do the same but as this is at ground level means the new growth will attract hungry mouths. Pollarding allows livestock into the wood to feed, whereas coppicing would mean they would need to be excluded with a fence.

Old Wives Lees began as *eald wudu* 'the old wood' and not until the seventeenth century did we see the addition of *laes* 'woodland pasture'. Solesdane Wood and Soleshill can be traced back to 1327 when the name is recorded simply as Sole. This is derived from *sol* or 'muddy pool', to which 'valley' and 'hill' were added comparatively recently. Wales Wood can be traced to the late fourteenth century when this place was home to the family of John Waleys.

Chillenden

Found as Ciollandene in the early ninth century and as Cilledene in Domesday. Here the Saxon or Jute personal name precedes Old English *denu*, telling us this was known as 'the valley of a man called Ciolla'.

Chislet

A name which possibly comes from Old English *cistelet*, which would refer to 'the chestnut copse'. There are two other suggested definitions, *cist laet* 'the water conduit of the chestnut tree', or *cyst laet* and 'the water conduit container'. The name is found as Cistelet in 605 and again in Domesday.

Boardedhouse Farm is self-explanatory, however with this being found as early as the fifteenth century it does show how unusual such a construction would have been at the time. From Middle English *boian tone* comes 'the farmstead of the young men', the name now transferred to Boyden Hill. The marshland around Chitty gave it the name of 'the island or dry land associated with a man called Citta', *eg* following a Saxon or Jute personal name. Grove Court is derived from Old English *graf*, 'a grove of trees'.

Nethergong Farm has undergone minor changes but enough to change the meaning. It began as Ouergange in 1278 and by 1393 became Overgonge. Here we see the common migration of the final 'n' in *atten* to the beginning of the following word as it begins with a vowel. Hence what began as 'the riverbank way', and a name suggesting an elevated position, has evolved to appear to refer to 'the nether or lower path or way'. What began as a place alongside the river was later also applied to the stream which drained the land here although there is no written evidence of the stream name being used for at least two centuries.

Without early forms we would define Port Farm as 'market', for this was the original meaning of Old English *port*. However from the thirteenth century we find Fayreport and Faireport, both showing this to be *faeger port* or 'the beautiful portal or entrance', its location suggests this referred to a small harbour on the Stour. Walmer's Hill remembers the family of John Wallemere, resident here in 1332.

Cliffe, West

Another from Old English *clif* and meaning the '(place at) the cliff or bank', this place is recorded as Wesclive in Domesday. The addition is self-explanatory.

Cliftonville

A nineteenth century resort with a name created to suggest its coastal location.

Coldred

Domesday lists this name as Colvet in 1086. Here is a name which describes a typical scene found across the land until at least the sixteenth century. Derived from Old English *col ryde*, this describes 'the clearing where charcoal is made'. The charcoal burners craft was highly skilled, having to cut fairly uniform lengths of wood, to stack them to form a large dome, cover with earth and burn slowly so that the water and impurities were removed but the slow burning charcoal remained. Clearly the skill was to allow sufficient combustion to achieve a result, without the whole pile being consumed by flame. To achieve this the man would have to sit for two, three or more days and keep an eye on his work. To say this was boring is an understatement and hence they took to sitting on a stool with just one leg. Such a difficult balancing act meant if they should succumb to sleep they would fall from their seat and be awoken. It is from this we get the expression "to drop off" meaning to fall asleep.

Locally we find Newsole Farm, an interesting name which comes from Old English *niwe sol* and telling of 'the new muddy pool'. It is difficult to explain how this muddy pool appeared without another factor being involved – a new route for carts, movement of livestock, nearby cultivation or irrigation. From the perspective of defining place names we would normally expect the reason for the creation of the muddy pool to be the basis for the name.

It is quite reasonable to think the name indicates this muddy place was troublesome and an inconvenience, although it does raise the question as to any earlier name for the place.

Wadholt Wood is found as Wlatenholte in 1270 and as Wadynholte in 1327, both clear evidence how unreliable the Domesday record of proper names can be when it gives this as Platenovt. The true origin here is a personal name followed by Old English *holt* and describing 'the thicket of a man called Wlata'.

Cranbrook

The earliest surviving record of this name is as Cranebroca from the eleventh century. Here is 'the brook frequented by cranes or herons', derived from Old English *cran broc*. The name of the River Crane is discussed under its own entry.

Cranbrook's attractive sign explains the meaning of the place name

Angley Wood has been suggested as referring to a bend or angle in the river, which is topographically correct but almost certainly not toponomically accurate. Here we find Old English *angeling leah* or 'the woodland clearing associated with the Angles', a name distinguishing them from the nearby Saxons.

Baker's Cross is held to be where Sir John Baker heard of the accession of Queen Elizabeth I. Some may have rejoiced at such news, however Sir John had arrived at Cranbrook to preside over a trial of Protestants, he being a renowned anti-Protestant during the reign of Mary I, a devout Catholic also known as Bloody Mary.

Barnden is derived from 'the *denn* or woodland pasture of a man called Beorn'. Birches Wood takes its name from the family of John Birchett, here by 1647. Recorded since the thirteenth century, Camden Hill is from *cumb denn* 'the woodland pasture in a valley'. Old Conghurst and Little Conghurst share a name from Old English *cung hyrst* 'the wooded hill overlooking the bend (in a stream)'. Friezley appears as Frithesleah in 804 and Fretheslee in 1250, this being 'the woodland clearing of a man called Frithu', the Saxon or Jute personal name suffixed by *leah*.

Copden Wood has been seen since the end of the fifteenth century, this coming from 'the *tun* or farmstead of a man called Coppa'. Couchman's Wood derives its name from the fifteenth century, when the family of William Cowcheman were living here. Flishinghurst comes from Old English *inga hyrst*, this telling us it was 'the wooded hill of the family or followers of a man called Plussa'. The change from an initial 'P' to 'F' can only have been down to it being copied incorrectly, by no means a rare occurrence, for the early forms give Plussinghirst in 804, Plussinghurst in 1278, and Pulsingherst in 1313. Further evidence of the error is found in the, now lost, place name of Plusshenden.

Glassenbury is from 'the stronghold of the Glastingas', the tribal name seemingly describing 'the resplendent ones' and thought to be because their weapons were unusually brightly polished. The nearby hamlet of Hemsted is recorded as Haemstede in 993, from Old English *ham stede* this is 'the place of the homestead'. Hensill comes from the Old English phrase *aet thaem hean gesellum*, referring to this being the '(place) at the high buildings'. Willsley Green can be traced to 1226 when we find Wiueleslegh, this being 'the *leah* or woodland clearing of a man called Wifel'.

Hurst was formerly known as Falconhurst. The present and original name come from *hyrst* meaning 'wooded hill'. During the thirteenth century Robert Falconarius held this manor, he being falconer to the king. Dating from 1258, the name of Marlings comes from the family of Thomas and John de Meling. Nichol's Wood is derived from a local family, William Nichole recorded here by 1278. From the family of William Sandre, here by 1381, comes the name of Saunders Wood. Trenley derives its name from *trind leah*, the name meaning 'the round woodland clearing' is recorded as Trindle in 1347.

The Peacock Inn features the image of a bird which allowed sign painters to show their talents. Yet the spectacular image of the male bird would equally have been chosen for a coat of arms. It should also be noted this was also a nickname applied to someone considered overly proud. Another bird gave its name to the Woodcock Inn, and thence to Woodcock Lane. Once the woodcock was considered a great delicacy and unlikely to have been served at such a pub.

Crane (River)

A river probably named from Old English *cran*, describing this as a place where 'cranes or herons' are seen. Although the earliest forms of this name date from the sixteenth century, this name may be much older and derive from a Celtic word related to Welsh *garan*. If the latter is correct and this is from Welsh *garan* it does not change the meaning for *cran* and *garan* have identical origins and meaning.

Crundale

Listed as Crundala around 1100, this comes from Old English *crundel* and refers to 'the chalk pit, a quarry'.

To the west is Anvil Green, which is recorded as Haugnefeld in 1240 and as Hannefeld in 1254. The present name is a corruption of the latter form and tells of 'the *feld* or open land of a man called Hagona'. Ashenfield Farm is found in Domesday as Esmerefel, this being 'the *feld* or open land of a man called

Aescmaer'. Winchcombe Farm appears in 825 as Wincelcumbe, this from Old English *wincel cumb* and speaking of 'the valley in the corner of land'.

Barton Wood took the name of the original settlement here, Old English *bere tun* meaning 'barley farmstead'. Trimworth Manor features a Saxon personal name and Old English *worth* to tell of 'the enclosure of a man called Dreama'. Viney's Wood comes from Old English *fin haeg* and speaks of 'the enclosure near a heap of wood'.

The pub known as the Compasses takes the image from the coat of arms of the Masons. Workers in stone will have been honoured here for the largest building in the city was built thanks to their skills.

CHAPTER FOUR – D

Deal

Records of this name include Addelam in 1086 and as Dela in 1158. This name comes from Old English *dael* and refers to 'the hollow or valley'.

How Bridge crosses a small tributary, its name coming from Old English *hoh* 'the spur of land' here. Walmer is recorded as Walemere in 1087, a name from *weala mere* and a reference to 'the pool of the Britons'.

The Yew Tree is an almost predictable pub name, for it features a tree, and such is common to many pub names as it makes for a recognisable image and a real life marker. Furthermore with it being a common sight in churchyards and the church being an important landholder, it would be easy to see both tree and pub in the same area. No need to explain why the place name appears in the Deal Hoy pub, only need say a 'hoy' is a small sailing vessel once used to ferry passengers and goods short distances along the coast.

The Saracens Head was once a far more common pub name. It is derived from the image adopted by the many landowners whose ancestors journeyed to the Holy Land on one of the Crusades. At the Lord Clyde we see a name commemorating Sir Colin Campbell, who was instrumental in the relief of the siege at Lucknow in India. In the Green Beret we see a pub named for the Royal Marine Commandos. Augustus, Viscount Keppel is remembered by the Admiral Keppel public house, he commanded the British fleets in several important victories over the French in the late eighteenth century.

At the Five Ringers the invitation is specifically aimed at the campanologists or bell ringers from the nearby church where, at least when the pub was named, there were five bells and bell ropes.

Denge Marsh

Listed as Dengemersc in 774, this probably comes from Old English *denu ge mersc* and speaks of 'the marshland of the valley district'. However there are some suggestions of this being Old English *dyncge* meaning 'the marsh with manured land'.

Denton

This is a common place name which is most often, as here, derived from Old English *denu tun* and refer to 'the farmstead in the valley'. This name is recorded as Denetun in 799 and as Danetone in 1086.

From Old English *gat hyrst* meaning 'the wooded where goats are grazed' comes Gatteridge Farm, the name found as Gatherste in 1304, Gateherst in 1444, and in the modern form for the first time in 1655. Islingham Farm is found from the eighth century, this starting out as 'the homestead of the family or followers of a man called Esla'. Lodgelees Farm took its name from Lodgelees House, which must have taken its name from 'the lodge of the meadow' and features the suffix *laes*.

Maydeken can be traced back to the family of Robert Maidekyn and Clement Madekyn, both of whom are recorded in a document dated 1327 and who owned the land around here until it was taken by Henry VI. With Old English *ing tun* following a personal name, Tappington Farm refers to itself as 'the farmstead associated with a man called Taeppa'.

In the Jackdaw we have a pub name derived from a book entitled the *Ingoldsby Legends*. Written by Richard Harris Barham, former vicar of Denton, one of the tales speaks of the Jackdaw of Rheims.

Ditton

Records of this name include Dictun in the tenth century and as Dictune in the Domesday survey of 1086. This name is derived from Old English *dic tun* and describes 'the farmstead by a ditch or dyke'.

Doddington

A place name featuring a Saxon or Jute personal name and Old English *ing tun*, this name can be defined as 'the farmstead associated with a man called Dudda or Dodda'. The name is recorded as Duddingtun at the end of the eleventh century.

Bistock is of Middle English origin, where *boian stoc* tells of 'the place of the young men'. Frangbury is listed as Fraungeber in 1327, Fronckebery in 1338, and Frankebury in 1348 and describes 'the fortified place of a man called Fronca'. Oakenpole Wood is from *acen* falod and talks of 'the fold found near oak trees', the name recorded as Okenefold in 1220.

Dover

Probably still the nation's most famous port, the traditional gateway to continental Europe, this name appears as Dubris in the fourth century, as Dofrus at the end of the seventh century, and as Dovere in the Domesday record. This is an old river name, the River Dour being a Celtic river name from *dubras* and meaning quite simply 'the waters'.

Aycliff has no early forms, however this undoubtedly comes from Old English *heah clif* 'the high cliff', a perfect description of its location. Chilverton Elms is a minor name reminding us this was 'the farmstead of a woman called Ceolwaru', with the later addition of the tree. Records of Coxhill are found from the early fourteenth century, this being from *cocces hyll* or 'the hill frequented by cock birds'.

Crabble Hill takes the name of an old place name, a name referring to anything but a hill. Recorded as Crabbehole in 1227, this comes from Old

English *crabba hol* and describes 'the hole or hollow in which crabs are found'. Domesday's listing of Feringelai shows Farthingloe is from Old English *fearn inga hlaew* speaking of 'the hill or mound of the dwellers at the ferns'. The name of Maxton, the modern form as early as 1242, refers to 'the *tun* or farmstead of a man called Maca'.

Shooter's Hill is derived from Middle English *scheteres helde* meaning literally 'the slope of the shooters' and a reference to a place where archers practised.

Temple Ewell is seen as simply Ewelle in 959, the addition not seen until the thirteenth century, an indication this manor was held by the Knights Templar. The basic name comes from Old English *aewiell* or 'spring'. Underdown Road can be traced to a minor place name listed as Vnderdoun in 1292, this speaking of its location 'under or below the hill'. Famous for its white cliffs one is known as Shakespeare Cliff, so named as it is said to be described in the play *King Lear* and where Edgar and the Earl of Gloucester converse. On the subject of Shakespeare, one local pub is named from one of his most popular characters. In three of his plays we find the obese, lying, practical joker named Falstaff, while the public house takes his full name of Sir John Falstaff.

In the First and Last we see a pub name which was always used to describe the pub on the outskirts of a parish, thus the first when approaching and the last when leaving a place. In Dover this is taken to the extreme as it can boast to be the first and last in the country. The coastal location is also seen in the Coastguard. In the Butchers Arms we see evidence that early landlords were not limited to serving drink but also earned money by turning their hands to another trade. Meat was an obvious choice for the cellars were the perfect place to keep it cool in the days before refrigeration.

The Marquis of Granby is evidence this was one of the many taverns run by officers and men who had served under John Manners, Marquis of Granby. His popularity among his men meant he commanded tremendous loyalty from them and he rewarded that loyalty on their retirement by setting them up in business for which they named the pub in his honour. It was not until the end of the twentieth century the Louis Armstrong was named, for the new landlord brought his own jazz band to play here.

At the Admiral Harvey we find a pub named after Sir Eliab Harvey, whose naval career culminated in his command of the *Fighting Temeraire* at the Battle of Trafalgar. Outside the Three Cups the sign depicts an image taken from the arms of the Worshipful Company of Salters, those who deal in this valuable commodity. In the Gate Inn we have a name which probably dates from an early place name and when *geat* referred to a 'gap' or 'way'.

Few pubs can be named for a scene in a play but this is the case with the King Lear. As the last scenes are played out the eponymous character is brought to Dover by the Earl of Kent where he meets Cordelia, the only one of his three daughters to remain loyal throughout. The Boar's Head is probably religious, specifically a celebration of the birth of Christ in the splendid centrepiece of the table with the sign dating from the fourteenth century.

Dungeness

Listed as Dengenesse in 1335, this name describes 'the headland near Denge Marsh'. Here the name of the marsh is suffixed by Old English *naess* meaning 'promontory', with Denge Marsh discussed under its own entry.

Dunkirk

Unlike other examples of this name in England, there is no suggestion of this being a remoteness name, nor is it suggesting this was a place of turmoil like its French namesake. The Kent example is first recorded in 1790 and was transferred from the French town of Dunkerque, itself derived from Flemish *duine kerk* 'the church on the dunes' and one which was built here in the seventh century and dedicated to St Eloi.

Bossenden Farm gets its name from 'the narrow valley of a man called Bosing'. Potters Corner is first found in 1620, named after the family who owned the land here. Winterbourne is a more common name in the south of England, this referring to 'the stream which only flows in winter' when increased precipitation produced sufficient water.

Dymchurch

Recorded as Deman circe at the end of the eleventh century, this name comes from Old English *dema cirice* and describes itself as 'the church of the judge'.

Jesson Farm can be traced to the thirteenth century as 'the farmstead of a man called Geffrey'. Orgarswick is a local name found as Ordgaresuuice at the end of the eleventh century and named from 'the specialised or dairy farm of a man called Ordgar'.

CHAPTER FIVE – E

Easole Street

Listed as Oesewalum in 824 and as Eswalt in the Domesday record of 1086, this name is probably from Old English *es walu* and speaks of 'the ridge or banks associated with a god or gods'.

Eastchurch

Listed as Eastsyrse at the end of the eleventh century, this name comes from Old English *east cirice* and refers to 'the eastern church', that is west of another place.

Connetts is a minor name derived from the Colnette family who were living at Eastchurch by 1439. Cripps Farm is a reminder of the Crips family, known to be living here in the fourteenth century. The name of Kingsborough comes from Old English *cyninges beorg* and speaks of 'the barrow or mound of the king'. Oldhook, Newhook and Hook Quay are derived from *hoc*, Old English for 'hook or spur of land'.

Rayham is a local name recorded as Reyhamme in 1254. The date of this and other early records do not show if this is from Old English *aet paere eg hamme* or Middle English *atter eyhamme*. Yet both have ostensibly the same meaning in 'the island or dry land in the water meadow'. Rowetts can be traced to 1485, a time when the family of Thomas Rowhede were resident here. Shurland is from *scir land*, an Old English term for 'the agricultural land belonging to the shire' rather than under private ownership.

Eastling

Listed in the Domesday survey of 1086 as Eslinges, this name comes from a Saxon or Jute personal name and Old English *ingas*. Together this name tells of the '(place of) the family or followers of a man called Esla'.

Arnold's Oak Farm is a reminder of one Arnold de Bonoia, lord of this manor by 1278 and who is known as Arnald de Eseling shortly afterwards. Divan Wood remembers the family of John Dyve, here by 1242. Mincendane Wood is not recorded until comparatively modern times, indeed were it not for a solitary listing made while Henry VIII was on the throne, this name would doubtless remain unknown. Said document gives a mention of Mynchin Lane, itself derived from *myncen* and a reference to 'nuns'.

Tong comes from Old English *tang* and describes 'a tongue of land', this place recorded as Thoong in 1218.

East Rother (River)

An old Celtic name related to British *dubro* and Welsh *dwr* and describing 'the rapid one'.

Eastry

Records of this name include Eastorege in the ninth century and as Estrei in the Domesday record of 1086. Here the name comes from Old English *eastor ge* and describes 'the eastern district or region'.

Black Lane refers to the colour of the land crossed by this way and not, as is most often found, that this is overgrown and a shaded road. Hay Farm began as 'the hedged enclosure' from Old English *gehaeg*, the name has changed little since the original record of Haye in 1242. Hernden is found since the early thirteenth century, all pointing to an origin of 'the woodland pasture associated with a man called Hearda'.

Highborough Hill has no surviving early forms, hence it is impossible to know if this represents Old English *heah burh* 'the high fortified place' or, should this be from a later era, uses *burh* in the manorial sense, literally 'the high manor'. Rowling Court can be traced back to the thirteenth century where the forms show this to be a tribal reference to 'the people of Rola'. Selson is a local name from *scylf tun* to speak of 'the farmstead at the shelf of land'. Venson Farm is seen as Wendleston in 1254, this from Old English for 'the farmstead of the Vandal', not someone who wrecks everything but someone associated with the tribe best known for proving a rather large thorn in the side of Rome some years earlier.

Shingleton Farm is derived what must be a tribal nickname, for there seems no other explanation for a name meaning the '(place of) the shrunken or withered people'. Statenborough is undoubtedly from *aet thaem steapan beorge* or 'at the steep mound'. Yet the earliest surviving forms are as Estenberge in 1086, Stepenberga in 1100, and Stapinbergh in 1240, where the first element is not the same as the present form and with no records to show why or when the change took place.

Tickenhurst can be found in Domesday, where it is recorded as Ticheteste. This is from Old English *ticcena hyrst* and describes 'the wooded hill frequented by kids (young goats)'.

East Studdal

Recorded as Stodwalde in 1240, this comes from Old English *east stod weald* and describes 'the eastern marshy land overgrown with brushwood'.

Egerton

Recorded as Eardingtun around 1100 and as Egarditon in 1203, this comes from a Saxon or Jute personal name and Old English *ing tun* and describes 'the farmstead associated with a man called Ecgheard'.

Burscombe Farm is a local name describing 'the *cumb* or valley of a man called Brocc'. Creed Farm is recorded as Crede in 1292, a name from *cread* which is Old English for 'the place of weeds'. As most place names are coined by neighbours, it is clear they considered this farmer's efforts less than ideal. Field Farm is derived from Old English *filethe* meaning 'hay' and indicating somewhere such was stored.

From Old English *gaers teag*, 'the grassy enclosure' was recorded as Grasteghe in 1250, as Grauftegh in 1270, and on modern maps as Grafty Green. Kingsden Farm is first seen in the thirteenth century and comes from *cyninges denn* or 'the woodland pasture of the king'. Wanden comes from Old English *wann denn* and speaks of 'the dark woodland pasture', the colour most likely referring to the soil and/or vegetation rather than shade.

Elham

The Domesday record lists this name as Alham in 1086. This name comes from Old English *ael hamm* and describes 'the hemmed in land where eels are found'.

A lovely old fingerpost at Elham

Acrise is a local name from *ac hris* 'the brushwood by the oak trees'. Bladbean is a strange name from *blod beam* and describes 'the blood beam or tree', a name also seen in Redoack just half a mile away, both referring to the red oak *Quercus borealis*. Boyke Wood comes from *boian wic*, Middle English for 'the specialised farmstead of the young men' – the speciality almost always dairy produce. Wingmore appears as Wengeme in 1240 and as Wyngemere in 1357, this coming from Old English *inga mere* and a Saxon or Jute personal name and referring to 'the pool of the family or followers of a man called Wiga'.

Canter Wood is found as Kanteworth in 1240 and as Cantreworthe in 1313, this descibes 'the *worth* or enclosure of a woman called Caentwaru'. Clavertye Wood comes from Old English *claefre teag* 'the clover enclosure' and is first seen as Clavertegh in 1341. Exted comes from Old English *ac stede*, meaning 'the place of the oak tree' and recorded as Hacstede in the twelfth century, Acstede in 1292, and as Axstede in 1347.

Winterage Farm is recorded as Wintringe in 1240 and most likely comes from a Saxon personal name and Old English *ingas* which tells of the '(place) of the family or followers of a man called Wintra'. However there are some schools of thought which suggest this could be a nickname for early Viking invaders and refers to a place where they made a temporary encampment and over wintered here.

Frarne Park represents the remaining part of a monastic estate known as Monkenlands in 1535, as evidenced by the Middle English *frerene* meaning 'pertaining to the friars'. Greenacre Farm is recorded from the thirteenth century, although the name is much older and from *grimes aecer* 'the cultivated land associated with a goblin or demon'. Ladwood took its name from a family, their surname either Lad or Ladde, who were here in the thirteenth and fourteenth centuries.

Oxroad Farm is first recorded in 1242, appearing as Oxrode and showing this to be from *oxa rod* and telling us we are standing at 'the clearing where oxen are penned'. Little Shuttlesfield is thought to take its name from a nearby stream once known as the Shuttle. Old English *sceathel* speaks of 'the divider' and thus the place name 'the open land near the stream known as the divider'. Seen as Wealdingworth in 1278, a name from Old English *wealdinga worth* and speaking of 'the enclosure of the dwellers of the now cleared forest', present-day maps show this as Wheelbarrow Town. The suffix barrow is down to the a change from *worth* to *burgh* in later centuries, the latter used in the sense of 'manor', while the later addition of Town is from Middle English *tone*, itself from Old English *tun* or 'farmstead'.

Pubs here include the New Inn, a name which may be self-explanatory but also tells us this was not the original pub here. Often such premises have not been 'new' for decades and, in retrospect, perhaps should have been named the Newer Inn. The Palm Tree Inn has not been new for a couple of centuries, for it was named by a crewman on the voyages of Captain James Cook, who would have seen many a palm tree on his voyages.

Elmstead

Recorded as Elmanstede in 811, this name comes from Old English *elm ham stede* and describes 'the homestead by the elm trees'.

Elmsted village sign

Ittinge Farm is found as Edinges in 1240 and as Edinge in 1278, a name describing 'the place of people of Eade'. Maxted Street is misnomer, for it comes from Old English *meox stede* and describes 'the place where dung is heaped'. Unlikely to be anything to do with fertilising the land, this is probably a dump or midden. Spong Farm is recorded as Sponge in 1334, although the name is from *spong* a dialect term for 'a long narrow field'.

Elmstone

Documented as Ailmereston in 1203, this comes from a Saxon or Jute personal name and Old English *tun* and describes 'the farmstead of a man called Aethelmaer'.

Cockerils Wood takes its name from the family of William Cokerel, recorded here by 1327. Preston is one of the most common names in England, from Old English *preost tun* this is 'the farmstead of the priests'. Sweech Farm can be traced to the fourteenth century and comes from *swice* Old English meaning 'trap, snare'.

Ewell (Minnis & Temple)

Two places which share a common origin from Old English *aewell* and refers to the '(place of) the source of the river'. These are recorded as Aewille in 772 and as Ewelle in 1086, the additions here come from *maennes* meaning 'common land' and show possession by the Knights Templar from the twelfth century respectively.

CHAPTER SIX – F

Faversham

Records of this name include Fefresham in 811 and as Faversham in 1086. Derived from Old English *faefer ham*, this name speaks of 'the homestead of the smith or metalworker'.

Bayfield is from *beg feld* 'the open land where prickly shrubs grow'. Brenley Corner is from *brand leah*, Old English for 'the woodland clearing cleared of undergrowth by burning'. Note this is a natural clearing where brambles and the like have grown and this is what has been burned, not the trees themselves.

Brogdale is found as Brokesdale in 1240, from Old English *broc dael* 'the marshy valley'. The 'agricultural land granted by charter' is today seen as Buckland, this derived from Old English *boc land*. The earliest surviving record of Davington is from the eleventh century as Denentune. This features Old English *denu tun* and speaks of 'the farmstead in a valley'. Kingsfield appears as Kingesland in 1209, the name describing 'the open land of the king' today and 'the agricultural land of the king' in the early thirteenth century – both the same feature.

Luddenham is recorded as Dodeham in Domesday, the modern name coming from 'the homestead of a man called Luda'. River is not a river but a settlement name, however it is named after a river and the meaning is obvious. Stubblefield means the same today as it did in Saxons times, however it was not always known by this name but was called Otershe. This is from *ate ersc* and describe 'the field of stubble where oats have been grown'. Wren's Hill gets its name from the family of Thomas Wrenne, here by 1260.

Syndale Bottom is recorded as Syndan in 1196, this early form showing it was from Old English *sin denu* meaning 'the huge valley'. The addition refers to a depression or hollow and is not seen until comparatively recently. In North Wilderton, South Wilderton and Wilderton Wood we have three names with a single source, this can be traced back to just one family and one man, for this is 'the *tun* or farmstead of a man called Wilhere'.

Public houses include the Chimney Boy, an image so often seen in the novels of Charles Dickens. These young boys were given the awful job of sweeping the chimneys by climbing up the inside. Eventually they grew too big to do the job and were cast aside, by which time their health will have suffered, assuming they had avoided serious injury or even death. Legislation was passed in the form of the Chimney Sweepers Act of 1788 which hoped to regulate the use of young boys by increasing the minimum age of an apprentice to eight years and only then with the consent of his parents, that the master should provide adequate clothing and living conditions, and allow the boy to attend church every Sunday. Despite this, and the introduction of the Humane Sweeping Machine in 1803, it was not until 1875, with the suffocation of George Brewster in a chimney at Fulford Hospital, that Lord Shaftesbury managed to change the law and effectively ban the use of boys.

The Elephant has been explained a number of ways, and while doubtless some were taken from heraldry and even suggesting one could get 'elephant's trunk' (drunk), most simply used the easily recognised image for their sign.

Finglesham

Recorded as Thenglesham in 832 and as Flengvessam in 1086, this has two possible origins. If this is from Old English *thengel ham* this would speak of 'the village of the prince', however should the first element be a Saxon or Jute personal name then this would be 'the homestead of a man called Thengel'.

Folkestone

Listed as Folcanstan at the end of the seventh century and as Fulcheston in the Domesday record of 1086, this name comes from Old English *stan* with a Saxon or Jute personal name telling us it was 'the stone of a man called Folca'. Whilst the stone vanished centuries ago, we do know the reason for its importance as this was documented as being a meeting place for the hundred, a Saxon administrative region.

Cheriton is 'the farmstead with or by a church', from Old English *cirice tun* and recorded in the twelfth century as Ciriceton. Coolinge comes from Old English *corningas* or 'the place of the corn people'. Morehall has been seen since the thirteenth century, an obvious name meaning 'the hall on the moor' from *mor heal*. Shorncliffe's early forms show this to be 'the cut cliff' and a reference to the two streams which have eroded their cuts into the cliff. Wingate Hill comes from *wind geat* and describes 'the windy gap (in the cliffs)'.

Danton Farm tells us it was known as 'the farmstead of the dalesman', the name recorded as Dalmynton in 1327 and as Damentone in 1348. Underhill is found as Hondheld in 1185, from *under hielde* or 'under the slope'. Foord, a hamlet near the coast, is a self-explanatory name recorded as Fforde in 1357. Walton Farm is situated at the former 'farmstead of the Britons', Old English *weala* meaning 'foreigner' and eventually used as the English name for the Welsh. Holywell is found since the thirteenth century, a name from *halig wielle* and telling of 'the holy spring'. Horn Street takes the name of 'the corner of land' from Old English *hyrne*.

Pub names begin with a delightful story concerning the origin of the Cherry Pickers. As usual the sign painter has been overly imaginative and not researched the origins. Here the image shows two magpies collecting the fruit from the tree. Opting for this species of bird seems an arbitrary choice, although we should be thankful the name is not illustrated by the mobile platform! The name was suggested by the 11th Hussars being based at Shorncliffe, they being nicknamed the Cherry Pickers. Best remembered for their bravery in the Charge of the Light Brigade at Balaclava, they earned this nickname during another war, yet another with the French, and being the victims of a surprise attack by French cavalry – the Hussars being completely unprepared as weapons were a hindrance when picking cherries.

The Martello is a pub name which would never have been named such had the correct term been used. During the early nineteenth century the threat of attack by the French under Napoleon meant the appearance of large towers along the coastline facing the Continent. Circular, with stone-built and extremely thick walls, these were miniature forts which could signal to one another and raise the alarm as soon as the enemy was sighted. The idea was copied from a tower in Corsica, captured by them in 1794. Sadly the name was always said to be Martello, when the original was found at Cape Martella. Named at the same time the British Lion public house symbolised the strength and fight of the British armies.

In the Master Brewer we not only find a name where the product is advertised, but also a claim of this product being an excellent beer. Today the microbreweries are sought by real ale enthusiasts, this name dates from a time when many pubs brewed their own for sale solely on their premises. No doubting the message of the Thistle and Shamrock, for these represent Scotland and Ireland respectively. At the Pullman public house we see a link to the railway through the name of George Pullman of Chicago. He gave his name to the style of railway coach which meant comfort and style, exactly the message at the pub.

Chambers is an alternative name showing it was not only in an area where the legal profession dominated but hoped to attract them as customers. Nailbox may point to another trade but not the obvious one of carpentry. This is a slang term once used in some publishing circles for critical written comments.

Fordwich

Listed as Fordeuuicum in 675 and as Forewic in 1086, this name is derived from Old English *ford wic* and describes 'the specialist farmstead at the ford'. Normally the *wic* would be seen as specialising in dairy produce, however here the records show good reason to believe this was a trading settlement.

The George and Dragon public house was named after the patron saint of England and his most famous action. Not only is St George the patron saint of England but also the Netherlands, Aragon, Bavaria, Germany, Lebanon, Bulgaria, Catalonia, Ethiopia, Georgia, Gozo, Greece, Malta, Sicily, Palestine, Portugal, Slovenia, Amersfoort, Beirut, Cappadocia, Constantinople, Ferrara, Genoa, Haldern, Heide, Lod, Modica, Moscow, Piran, Ptuj, Reggio Calabria, agricultural workers, archers, armourers, butchers, cavalry, chivalry, Corinthians (a Brazilian football club), Crusaders, equestrians, farmers, field workers, herpes, horsemen, horses, husbandmen, knights, lepers, leprosy, the Order of the Garter, Palestinian Christians, plague, Portuguese Army, Portuguese Navy, riders, Romani people, saddle makers, scouts, sheep, shepherds, skin diseases, soldiers, syphilis, and Teutonic knights.

Foreland (North & South)

Two places sharing a common origin in Old English *fore land* and speaking of 'the promontory'. The earliest record dates from 1326 as Forland and, while the two manors are otherwise unconnected, they do have identical meaning and are named to distinguish between the two.

Frinsted

Domesday lists this as Fredenestede in 1086, the name derived from Old English *frithen stede* and speaking of 'the place of protection'.

Browning's Wood is named after the family who were here by the thirteenth century, although the spelling of the name was then seen as Broning, Bruning and Brounyng. Madams Court was known as Meriams Court in 1782, with five centuries earlier the record of the family of Radolphus de Mayham being the origin of the name. Wrinsted Court is recorded as Wrensted in 1111, this from Old English *wrenna stede* and meaning 'the place frequented by wrens'.

Frittenden

Documented as Friththingden in the ninth century, this name features a Saxon or Jute personal name and Old English *ing denu* and describes 'the valley associated with a man called Frith'.

Frittenden village sign

Brissenden has been recorded since the fourteenth century, the name describing 'the woodland pasture of a man called Breosa'. Knox Bridge refers to the crossing, its name, ultimately from *cnoll* 'hillock', was brought here as a family name. Sinkhurst Green is probably from Middle English *sinke hurst*, a name describing 'the wooded hill by the low-lying land'. Little Wadd Farm was recorded as simply Wold in 1240, from *weald* this is one of several places named for being on 'the forest area now cleared'. Whitsunden is found in a document dating from 1202 as Wichenden. This features an Old English suffix of *denn* with a Saxon personal name and speaking of 'the woodland clearing of a man called Wicca'.

Pubs include the Bell and Jorrocks, which were once two quite separate premises. He former name is easy to see as showing this was land belonging to the church, while John Jorrocks is a character created by nineteenth century novelist Robert Surtees. Jorrocks was a Cockney grocer, whose unique personality leads him into a series of adventures.

CHAPTER SEVEN – G

Godmersham

Listed as Godmeresham in 822 and as Gomersham in 1086, comes from a Saxon or Jute personal name and Old English *ham* and telling of 'the homestead of a man called Godmaer'.

Left: Godmersham church of St Lawrence the Martyr
Right: Great Stour at Godmersham

Bilting describes the '(place) of the people of a man called Belt'. Found as Beltinge in 1272, that same document also gives the name as Beltesburne showing the stream was also associated with this man. Buckwell comes from *bucca wielle* or 'the spring frequented by bucks'. Coneyearth Wood is a corruption of 'the enclosure of a man called Caena', the modern form clearly shows this was mistakenly believed to refer to rabbits.

Popestreet Farm dates from the early fourteenth century when the family of Peter and William le Pope were recorded as resident here. Purr Wood comes from Old English *pur wudu*, literally 'unblemished wood' which is understood to be a wood which has never been harvested or used previously. Temple Hill is seen simply as Temple in a document of 1313, the very obvious hill was then a manor held by the Knights Templar.

Goodnestone

One of two places in the county of this name, this example is near Aylesham and is listed as Godwineston in 1196. Derived from a Saxon or Jute personal name and Old English *tun* and referring to 'the farmstead of a man called Godwine'.

Uffington Farm is seen as Vffingtun in 1226, this being from Old English *ing tun* and a Saxon personal name telling of 'the farmstead associated with a man called Uffa'.

Goodnestone

The second place of this name in Kent, this example near Faversham is listed as Godwineston in 1208. Again the name comes from a Saxon or Jute personal name and Old English *tun* and refers to 'the farmstead of a man called Godwine'. There is no suggestion this is the same person named in the previous entry, quite simply this is a common personal name.

Gore

The earliest record of this name, the document is dated 1198, is exactly the same as today. This is from Old English *gara* which refers to the '(place at) the triangular area of ground'.

Graveney

Recorded as Grafonaea during the ninth century, this name comes from Old English *grafa ea* and refers to the '(place at) the ditch stream'.

To the east we find the names of Nagden and Nagden Marshes, a name found from the sixteenth century. This is derived from Old English *nag dun* and speaks of 'the small or sturdy hill'. The Four Horseshoes is a reminder of days when the local inn and the blacksmith together offered the traveller similar services to today's motorway service station. The name is best seen as a question and, if the answer is negative, the solution is here.

Greatstone-on-Sea

Here is a recent name not seen before 1801 when it appeared as Great Stone, a name given to a rocky headland here before erosion changed the coastline.

Guston

Listed as Gocistone in 1086, this name comes from a Saxon or Jute personal name and Old English *tun* and tells us this place was 'the farmstead of a man called Guthsige'.

Poison Down and Poison Wood have few records and yet the only origin seems to be that shared by the now lost place name of Pising. If these do represent a corrupted version, then this represents Old English *pysingas* and describes the place of 'the short fat people'.

The local pub is the Chance Inn. It was named by the first landlord, a farmer who converted his cottage into a pub. This was seen as taking a chance by villagers and the name was coined.

CHAPTER EIGHT - H

Hackington

A place name listed as Latintone in the Domesday record of 1086, as Hakinton in 1186, and as Hakington in 1226. This is Old English *tun* following a Saxon personal name to describe 'the farmstead of a man called Hacca'.

Locally we find Brotherhood Farm, so called because it was once the property of the hospitals of St Nicholas and St John, both run by monks. The name of Cheesecourt Gate is a reminder of the family of Thomas Chesman, here before 1327. Daw's Wood was associated with the family of Thomas Dawe by 1327. Hales Place is named after Sir Edward Hales, resident here in the eighteenth century.

Honey Wood tells everyone this is 'the wood where bees are known or encouraged to nest', honey being a vital ingredient in Saxon life. With records dating back to the ninth century, Shelford Farm began as 'the farmstead of a man called Scufel'.

Halstow (High & Lower)

A name derived from Old English *halig stow* and telling us it was 'the holy place'. The additions are self-explanatory, with High Halstow recorded as Halgesto around 1100, around the same time as Lower Halstow appears as Halgastaw.

Callum Hill appears in a document dated 1598 as Horsham Croft alias Callums, the present name that of the landowner, the former describing 'the hemmed-in land where horses are reared'. Dalham comes from Old English *hael ham*, recorded as Daelham in 980, Delham in 1246, and Dalham in 1332 and describing 'the homestead in the valley'.

To the west is Fenn Street, recorded since the thirteenth century and derived from Old English *fenn* with a later addition showing this was 'the road leading across the fenland'. St Mary's Hoo takes the dedication of the church and adds it to Old English *hoh* or 'spur of land'. Sharnal Street comes from *scearn wielle* or 'the spring near a dung heap or midden', a warning name recorded since the fourteenth century.

Ham

Found in the Domesday record of 1086 as Hama, this does not come from Old English *ham* meaning 'homestead' but from *hamm*. It is often difficult to distinguish between these two Old English words when used as a suffix, however when it stands alone, as here, the latter *hamm* is often difficult to define for it has so many different uses. The word is used to variously describe 'land in a river bend', 'land almost surrounded by the topography' or 'land with water on three sides'. Possibly the best way to describe it is as 'hemmed in land', not cut off but leaving only one clear way in or out.

Updown Farm is seen in a document dated 1206 as Uppedun, this from *uppe dun* and telling it is 'up on a hill'.

Harbledown

Listed as Herebolddune in 1175, this name comes from a Saxon or Jute personal name and Old English *dun* and describes 'the hill of a man called Herebeald'.

China Farm is a corruption of Cheney and taken from Cheney Court, home to the family of Robert Cheney in the late fifteenth century. Poldhurst Wood is found as Polre in 1292, this being from Old English *polra* and describing 'the wetland'.

In the Coach and Horses we have a name which was unknown before coaching days and effectively tells travellers it was a stop on the old coaching routes.

Harrietsham

Listed as Heriheardes in the tenth century and as Hariardesham in Domesday, this name has two possible origins. Either this features a Saxon or Jute personal name and Old English *hamm* and refers to 'the hemmed in land of a man called Heregeard', or is from *geard hamm* and describes 'the hemmed in land near army quarters'.

Chegworth is recorded as Chagewrth in 1201 and Chagworthe in 1253, this is from *ceacge worth* 'the enclosure among or near the gorse'. Delaware is named after a local family who appear in various records between 1199 and 1346 as De la Ware, La Ware, and De la Warre all meaning 'dweller by the weir'. Recorded as Godyntone in 1318, Goddington refers to 'the *tun* or farmstead of a man called Godda'.

Holm Mill comes from *hol myln*, a name seen from the thirteenth century although 'the mill of the hollow' probably existed well before then. Jone Wood can only be from the family of Thomas Jon, here by 1481. Stedehill Wood and Stede Court share an origin in the family of Roger Stede, here by 1450.

Hartley

One of two places in the county of this name, this example is near Cranbrook and was recorded as Heiratleag in 843. The name is derived from Old English *heorot leah* and refers to 'the woodland clearing frequented by harts or stags'.

Swattenden is a local name found from the thirteenth century and featuring a Saxon personal name and Old English *denn* and telling of 'the woodland pasture of a man called Swaethel'. Turnden has existed in its present form since 1254, this from Old English *tyrn denn* and 'the woodland pasture by a turn or bend (in the road)'.

Hartlip

A name from Old English *heorot hliep* and speaking of 'the gate or fence over which harts or stags can leap', this is found in a late eleventh century document as Heordlyp.

The local name of Matts Hill Farm can be traced back to as early as 1404 when the family of Richard and William Met were recorded here.

Left: Hartlip Church of St Michael and All Angels
Right: Hartlip sign

Hastingleigh

Records of this name include Haestingalege in 993 and as Hastingelai in 1086. This name comes from a Saxon or Jute personal name and Old English *inga leah* and refers to 'the woodland clearing of the family or followers of a man called Haesta'.

To the north we find Bavinge, found as Babinge in 1320 and describing the '(place) of the people of Babba'. Bodsham Green is found in 675 with Botdesham and as Boddesham in 811, this is 'the homestead of a man called Bodd'. Coombegrove Farm means exactly what it says 'the grove of trees in the valley'. English Wood is derived from the family of Emme, Robert and Thomas Englisshe, all recorded here in a document of 1348.

Wicken can be traced to the late eighteenth century when it was home to the family of Henry de Wicombe. Records of Lyddendane Farm can be traced back to the thirteenth century, where a Saxon personal name precedes Old English *dene* to refer to 'the valley of a man called Leofwynn'.

The Bowl Inn public house ....

.... and the road called Bowl Field are derived from a field name

Hawkhurst

Found as Hauekehurst in 1254, this name comes from Old English *hafoc hyrst* and describes 'the wooded hill frequented by hawks'.

Bare Tilt Farm is a modern name which was earlier found as Bertilth in 1284. This shows the early name describes 'the tilled land where barley grows', hence what began as a farming settlement and became a place name survives in the twenty-first century as the name of a farm. Barrett's Green takes the name of the family of Robert Barrett, here by 1349. Both are simple enough but not as obvious as that found just to the south and known as The Moor.

Basden Wood is first recorded as Badisdenne in 1253, a name meaning the *denn* or woodland pasture of a man called Baeddi'. Buckhurst is found in a document dated 1292 as Bokhurst, this being from *boc hyrst* 'the wooded hill of beech trees'. Cockshot is a fairly common minor name which refers to a glade where beaters would gather to scare birds into nets which had been placed at the far end and then gathered for the table, the term coming from

Old English *cocc sceat*. From Old English Lillesden describes 'the woodland pasture of a man called Lil'. Woodsden comes from Old English *wudures denn* and tells of 'the woodland pasture of the woodsman'.

Delmonden Green is recorded as Delmunden in 1206 and as Delmyndenne in 1347, from *daelmanna denn* this describes 'the woodland pasture of the dalesmen'. Four Throws to the south east comes from *feower treow* and does indeed point to a line of four trees used as a route marker. Gill's Green is named after the family of Matilda Gille, recorded here in a document dated 1327.

Hedgingford Wood is found as Heggingeworth in 1310, showing this name began describing 'the *worth* or enclosure of the family or followers of a man called Haecga'. The change from *worth* to *ford* occurs more often than we would expect.

Pipsden features the suffix *denn* which, with a Saxon personal name, describes 'the woodland pasture of a man called Pipp' and is recorded since at least the thirteenth century. Risden is recorded as Rissesdenn in 1240, this from *hris denn* and telling us it was 'the woodland pasture marked by brushwood'. Recorded as Sisele in 1230, Siseley features the suffix *leah* and describes 'the woodland clearing of a man called Sissa'. Slipmill is recorded since the fourteenth century, from *slaep myln* this is 'the mill by a slippery place'.

Soper's Lane can be traced back to the early sixteenth century when we also found Soper's Lane Farm. The name refers to a path crossing land owned by the Soper family. Similarly White's Wood recalls the middle of the thirteenth century when the family of Robert Wyte were resident here. It is not often we know something of the person who gave their name to a place, yet in the case of Winch's Plantation this is the case. In the eighteenth century Mr Winch was a factory owner employing a hundred or more workers producing worsted.

Pub names begin with the Kent Cricketer, the sport having been associated with the county for many years.

Hawkinge

Listed as Hauekinge in 1204, there are two possible origins for Old English *hafoc ing*. Either this can be taken literally as 'the place frequented by hawks', or perhaps the first element is used as a personal name in which case this would be 'Hafoc's place'.

Argrove Wood comes from Old English *ora graf* 'the woodland grove by a bank'. Drellingore is seen from the middle of the thirteenth century, speaking of 'the bank or slope of a man called Dylla'. The manorial name of Flegis Court is seen as Fleggescourt in 1346 and recalls the Fleg family, who certainly originated from Flegg in Norfolk.

Killing Wood is seen as Kellingesdene in 1263, this referring to 'the valley of the family or followers of a man called Cylli'. Palmer's Hill takes the name of William Palmer and his family, documented here in 1327. Pay Street was home to Henry Peys by 1348, the family well recorded in the area around Folkestone. Reinden Wood has changed little since its first record in 1262, this being from *raegena denn* and a reference to 'the woodland pasture frequented by does'.

Terlingham is recorded as Terlingeham in 1262 and as the modern form for the first time just eight years later. This comes from Old English *ham* and a tribal name referring literally to 'the homestead of the trillers'. The nickname is understood as showing they were known for their singing.

Headcorn

At the end of the eleventh century this name appears in a document as Hedekaruna. This is from a Saxon or Jute personal name and Old English *hruna* and speaks of 'the tree trunk of a man called Hydeca'. We also know this tree trunk would have been used as a footbridge and thus defining the name gives an impression of the place, a picture which would never have been produced on canvas.

Baker Lane is not directly from the trade but from a family whose ancestors earned their name from baking, this being associated with the family of Thomas Baker before 1472. Barling Green is a transferred name, although somehow

Barming became Barling over those few miles when first seen as Barmlyng in 1327 and clearly brought here by a settler. Wick Hill features the element *wic* or 'specialised farm', most often that speciality is dairy produce and here clearly one on a hill. Seen as Wideringdenn in 863 and as Wytheryndenne in 1292, the name of Witherden Farm speaks of 'the woodland pasture of a man called Wither'.

Bletchenden is seen as early as the eighth century, found as Blechinden and telling of 'the woodland pasture of a man called Blecca'. Glovers Bridge gets its name from the family with many representatives over the years, including Henry Glouere in 1327 and John and Richard Glover in 1450. Hawkenbury is found as Focgingabyra in 814, not unlike the original Old English *focginga burh* describing 'the fortified place of the dwellers of the bog'. This area is at the fork of low-lying land created by the union of the River Beult and one of its minor tributaries.

Kelsham is from Old English *hamm* and a personal name speaking of 'the water meadow of a man called Cylli'. Ramhurst is recorded as Ramherst in 1313 and in the modern form for the first time in 1347. This is from *ramm hyrst* or 'the wooded hill where rams are reared'. Southernden is found as Swytheryndenn in 1292, here a Saxon personal name is suffixed by Old English *denn* to refer to 'the woodland pasture of a man called Swithred'.

Herne (Herne Bay)

Found as Hyrnan around 1100, this name certainly comes from Old English *hyrne* and referring to the '(place at) the angle of corner of land'. Clearly the land name was later transferred to the nearby bay.

Locally we find Beltinge, which has not changed since a document of 1189 and refers to the '(place of) the people of a man called Belt'. Bishopstone is from Old English *biscopes tun* 'the farmstead of the bishop'. Blacksole Farm is from *blaec sol* 'the farm at the muddy pool'. Bleangate takes the name of the local hundred, this being the meeting place for this basic administrative region and describing 'the way to the Forest of Blean. Bullockstone is a hamlet about a mile west of the town, first recorded in 1348 it takes its name from the family of Boullynge and is thus 'the *tun* or farmstead of the Boullynges family'.

Cobblers Bridge Road is a reminder of this area being associated with the family of John and William Cobelot in the middle of the fourteenth century. Crowdown Wood took its name from 'the woodland pasture frequented by crows', from Old English *crawe denn* and listed as Crowdenne in 1485. Eddington is found as Edington in 1466 and Hedinton in 1506, this features a Saxon or Jute personal name with Old English *ing tun* and tells of 'the farmstead associated with a man called Edda'.

Gilling Drove, recorded as Gildenge in the thirteenth century, an old drovers route across land which had previously been associated with a tribe called the Gildingas. Recorded as Haneford in 1270 and derived from *hana ford*, the 'ford frequented by cock birds' has become the modern oxymoron of Henfoote Farm. Hicks Forstal Road is a reminder of the old place name found in 1313, this featuring the dialect word *forstal* in describing 'the land in front of farm buildings associated with the Hicke or Hycke family'.

Highstead, found from the thirteenth century, describes 'the high place' from *heah stede*. From Old English *halig beorg*, Hillborough probably describes 'the place of worship of the tumulus'. Hunters Forstal first appears as Hunter Street in the fifteenth century, the name is probably manorial with *forstal* a dialect word for 'land in front of farm buildings'. Maystreet takes the name of former resident William May, he is documented here in 1327. Northwood is self-explanatory and is found to the north of the parish.

Surviving records of Oxenden do not begin before 1790, however we are still fairly confident this represents *oxena denn* and tells of 'the woodland pasture for oxen'. Ridgeway Farm does indeed stand on a 'way on a ridge', the name recorded by the early thirteenth century and coming from Old English *hrycg weg*. Stud Hill does not show it was a former breeding establishment for horses but comes from an old resident. In 1327 William Stud was farming nearby, with Studhill first seen in 1473.

The earliest surviving record of Talmead comes from a document dated 1500. However the name is much older, coming from 'old mead or meadow'. That the name begins with a 'T' when it should clearly begin with the vowel is a common occurrence. Here the last letter of 'at' has migrated to the front of the following word. Thundersland Road can be traced to 1357 and the first reference to the family of Richard Thunder being in residence here.

Pubs begin with the Queen Victoria, another in the many examples which also makes our longest reigning monarch (to date) the woman with the most pubs named after her. Indeed this is not the only pub in Herne to be named after the queen, yet there would be little confusion unless you could see the sign outside the Bun Penny. This refers to the image of the queen on the penny black, the first stamp issued anywhere in the world, where she is wearing her hair in a bun.

In the Druids Head we have an indication this was a meeting house for the United Ancient Order of Druids, a friendly society founded in 1781. It seems obligatory to find almost half of all the coastal settlements in the land to have a pub named the Smugglers Inn. The slurred pronunciation of the place name seems to have given a name to the Heron public house. An excellent and likely unique variation on the common name of the plough is seen in the Share and Coulter, these are the two kinds of blades found on the plough.

Hernhill

Another name where the earliest record comes from around 1100, this time seen as Haranhylle. From Old English *har hyll* this name speaks of the '(place at) the grey hill'.

Locally we find Butler's Hill, the place named from the family of Bottrell, certainly here by 1540. Dargate is found as Deregate in 1275 and Dergate in 1458, this being from Old English *deor geat* 'the way or track used by wild animals'. Denly Hill is seen as Denelee in 1214, from Old English *denn leah* this is 'the pasture at the woodland clearing'. Nearby Dernstroude is from *denn steorfa*, literally the 'pasture of pestilence' and a reference to unproductive, possibly sour, soil. Waystreet Farm is recorded as La wey in 1240 and as simply Weye in 1327, this from Old English *weg* and telling of 'the track or road'.

In the pub name of the Four Horseshoes, we see an advertisement for the local blacksmith. The metalworker and the publican often went hand in hand to provide a service akin to that of the modern service station. It is easier to see as such if viewed as a question when, if the answer is negative, this place provides the solution.

High Halden

Recorded as Hadinwoldungdenne at the end of the eleventh century, this features a Saxon or Jute personal name and Old English *ing denn* and describes 'the woodland pasture associated with a man called Heathwald'. The modern form is much shorter but has gained an addition which most often describes its greater importance than its elevation.

Broombourne Farm comes from Old English *brun burna* 'the brown stream', the name seen as Brombourn in 1347 and Brunbourne in 1348. Butterfield Wood was named from 'the ford by the good pasture', although the origin is literally 'the butter ford' from *butere ford*. Egerden has changed little since the late thirteenth century, this describing 'the woodland pasture of a man called Eadgar'.

Harbourne House was built on land known as Hyringburne in 968, itself named after the small tributary of the Highknock Channel known as 'the stream of a man called Hyra'.

Mace Wood can be traced back to the seventeenth century when this was known as May's Wood, and thus a name first seen in 1609 when Reginald May is recorded here. Plurenden Farm and Plurenden Lane feature the suffix *denn* with a Saxon personal name and describe 'the woodland pasture of a man called Plera'.

Hinxhill

Documented as Haenostesyle around the end of the eleventh century, this may come from Old English *hengest hyll* and refer to 'the hill of the stallion' or perhaps the first element is a Saxon or Jute personal name and thus 'Hengest's hill'.

Fingerpost at Hinxhill

Hinxhill village sign

Locally we find Goodcheap Farm, a name recorded as Godchep in 1270 and derived from Old English *god ceap*. The literally meaning is 'good market', perhaps an indication this was known to consistently undercut competing markets on price.

Hoath

In the thirteenth century this name is recorded as La Hathe which, despite the French definite article, is from Old English *hath* and describes the '(place at) the heath'.

There is a local place name, that of Rushbourne, which is clearly derived from a stream name and yet no stream here is known by this name today nor is there any record of such. However the definition is easily seen as 'the stream where rushes grow' from Old English *risc burna*. The present name of that same watercourse is the Sarre, which must have been the original name for this is a Celtic or British river name related to Latin *serare* meaning 'to close or shut' and a reference to dams constructed along the stream.

In the Prince of Wales is a name most often referring to the future Edward VII, although as the title held by the eldest male child of the reigning monarch was first given to the future Edward II. There are two reasons for the son of Queen Victoria being the most common reference to the Prince of Wales. Firstly the number of pubs built during his time in office as the populations of our towns and cities exploded with the number of jobs created as a result of the Industrial Revolution. Secondly he held the title for 59 years and 45 days, longer than anyone in history. The present incumbent, Prince Charles, is due to equal this milestone on 9th September 2017, when he will be approaching 69 years of age and his mother over 91.

Hollingbourne

Records of this name include Holingeburna in the tenth century and as Holingeborne in Domesday, this name has two possible origins. Either this comes from a Saxon or Jute personal name and Old English *inga burna* and describes 'the stream of the family or followers of a man called Hola', or alternatively this is from *hol inga burna* and means 'the stream of the people in the hollow'.

Locally Elsfield is derived from Old English *feld* with a Middle English personal name and describes 'the open land of a man called Eli'. Greenway

Court is from *grene weg* 'the green or overgrown way or track'. Mordenden Wood has plenty of records but none earlier than the thirteenth century. This is probably a Saxon personal name with *ing tun* telling of 'the farmstead associated with a man called Moda'. Penn Court takes Old English *pund* to tell of 'the pound' or animal enclosure. The earliest record of Snarkhurst Wood dates from 1645 as Snockhurst. This is from *snoc*, an early dialect word, and *hyrst* to describe 'a wooded hill on a projecting point of land'.

In the Park Gate Inn is a reminder of the parkland estate, this standing at the entrance to same. Once sugar was sold wrapped in easily recognised conical shapes, which were known by the same name as the Sugar Loaves public house.

Hoo (St Mary's Hoo)

The basic name here comes from Old English *hoh* and describes the '(place at) the spur of land'. The addition for the second name is, not surprisingly, a reference to the dedication of the church. Records of these places include Hoge in 687, How in 1086, and Ho St Mary in 1272.

Beluncle is a local name derived from a former inhabitant, the family of William Beluncle recorded here by 1240.

Hothfield

Documented as Hathfelde at the end of the eleventh century, this comes from Old English *hath feld* and speaks of 'the heathy open land'.

Hothfield sign

Bockhanger is from *boc hangra*, 'the slope covered by beech trees'. Swinford Old Manor can still be seen as coming from *swina ford* or 'the ford where swine cross'. Yonsea is recorded as Tntesie in 1231 and as Jenteseye in 1334, an area of drier ground in marsh known as 'the island of a man called Gent'.

Hucking

Coming from Old English *ingas* and a Saxon or Jute personal name, this speaks of the '(place of) the family or followers of a man called Huca. The place appears as Hugginges in a document of 1195.

Local names include Rumsted Court, which has early records of Thrusted, Thromstede, as Thrumstede and Thrumstead in the thirteenth and fourteenth centuries. The modern form has lost the first two letters, for this comes from Old English *thrim stede* and meaning 'the cut off place' and understood as a where trees were reduced to stumps to leave an area where buildings could be constructed.

To recognise a pub name without having to add the word 'pub' is actually quite simple to achieve. For example, in mentioning the George and Dragon or the Dog and Doublet our thoughts instantly turn to the public house. Just by linking two seemingly unrelated words with 'and' points to a pub name. If that means using the two most important words beginning with the same letter so much the better, for alliteration is still employed by advertisers. No product is mentioned in the phrases "Flexible friend", "Tiger in your Tank", or "Clunk, Click" and yet those of a certain age will recognise Access credit card, Esso petrol, and the seatbelt campaign, respectively.

In the Hook and Hatchet we not only have a pub name which uses alliteration, has two seemingly unrelated elements, but which was chosen for a very specific reason. This part of the county has close associations with the Royal Navy and both hook and hatchet are found on the badge of a Chief Petty Officer Shipwright, probably an indication of an early landlord's previous career. These tools were not chosen arbitrarily, for they remind us the Royal Navy once had tree-felling rights and thus could obtain timber for building ships virtually anywhere they desired.

Hythe

Old English *hyth* is used to refer specifically to 'the landing place, harbour'. This name is found as Hede in the Domesday record of 1086.

Casebourne Wood is seen as Casinburnan in 811, from Old English *aet paem casing burnan* or the '(place) at the stream of the dweller in the hovel'. Scene Farm appears as Seende in 1292, showing this comes from Old English *sende* to tell of 'the sandy place'. Another name derived from the coastal location is Seabrook, a name recorded since at least the sixteenth century and a reference to the 'sea marsh' here.

The pub named the Three Mariners takes its name from a ballad by W. S. Gilbert. In *The Yarn of the Nancy Bell* the mariner sings about being captain, cook, mate, bo'sun, midshipman, and indeed the whole crew. His jolly song hides the awful truth, how after being shipwrecked this naval man only survived by eating his two colleagues. The product is advertised on the sign outside the Butt of Sherry.

The Britannia is really a patriotic name, one which refers to the female figure who took the Roman name for our islands when she posed for a coin struck in the seventeenth century. For years that same image has appeared on coins, she is based on Frances Stewart who was later made Duchess of Richmond although her title was probably down to her being one of the many mistresses of Charles II, rather than for posing for the coinage.

CHAPTER NINE – I

Ickham

A name recorded as Ioccham in 785 and as Gecham in the Domesday book of 1086. This comes from Old English *geoc ham* and refers to 'the homestead comprising a yoke of land'. A 'yoke' is generally said to be approximately fifty acres of land.

Appleton is listed as in Domesday as Apletone, a reminder this was once a separate settlement from *aeppel tun* referring to 'the farmstead by the orchard'. Cherville House comes from Old English *ceorla feld* 'the open land of the churls or peasants, found as Cherlefeld in 1270 and Cherlefelde in 1338. Hazeling Wood comes from *haesel ing* and speaking of the '(place) associated with hazel trees'.

Lackenden Wood puts *ing dael* after a personal name and speaks of 'the dale associated with a man called Luca'. Well Chapel is simple enough, the name from Old English *wielle* or Middle English *wielle* and telling of 'the spring'.

Isle of Harty

A name recorded in Domesday as Hertei, this comes from Old English *heorot eg* 'the island of the hart or stag'.

Isle of Oxney

With records of this name as Oxnaiea in 724, Oxeneya in 1212, and as Oxney for the first time in 1610, we have both written and etymology evidence this is a Saxon place name. Here is Old English *oxena eg* telling of 'the island of the oxen'. This is not all we have to show how closely the modern version is to the Saxon name. While *eg* or 'island' is often used it is not often we find a true island, the majority are simply where higher and therefore drier ground is found in what is mostly marshland. Here the island is formed by two branches of the River Rother. Also excavations by archaeologists have discovered this was a pagan site of worship, with artefacts depicting this as where oxen were sacrificed to their gods.

Isle of Sheppey

Listed as Scepeig at the end of the seventh century and as Scape in Domesday, this name comes from Old English *sceap eg* and refers to 'the island where sheep are kept'.

Ivychurch

Recorded as Iuecirce in the eleventh century, this is derived from *ifig cirice*, an Old English term for 'the church covered in ivy'.

Iwade

In 1179 this name is found as Ywada. Here is Old English *iw waed* and speaks of 'the crossing place where yew trees grow'. This would have led to the Isle of Sheppey.

Chetney is a reminder of the old Chetney Marshes, recorded as early as 1370. However the marsh was named from the island or dry land which was the ideal place to live, for this name comes from the suffix *eg* and describes 'the dry land in a marsh of a man called Ceatta'. Coleshall has been recorded since the fourteenth century, the name describing 'the *halh* or corner of land of a man called Col'.

CHAPTER TEN – K

Kenardington

Here a Saxon or Jute personal name precedes Old English *ing tun* and tells us of 'the farmstead associated with a man called Cyneheard'. The name appears as Kynardingtune in a document dating from the eleventh century.

Locally we find Benchill Farm, seen as Benteleye in 1301 and coming from *beonet leah* 'the woodland clearing where bent grass grows'. From this definition we can also deduce this area must have been fallow land which remained ungrazed, Grass is a wonderfully successful group of plants but none grow bent which tells us this place was allowed to grow untouched until the plants collapsed under their own weight. We can also deduce that these people must have earned their living in another way other than agriculture.

Denne's Wood gets its name from the family who included Richard, Robert and Walter Denis, recorded as living at Kenardington in a document dated 1327.

Kennington

The second place of this name in the county is near Ashford. This example having a different origin and meaning. Listed as Chenintune in the Domesday record of 1086, this features a Saxon personal name and Old English *ing tun* and describes 'the farmstead associated with a man called Cena'.

To the north is the hamlet of Eastwell which, as the name suggests, is 'the eastern spring' from *east wielle*. Ulley Farm appears as Wlleg in 1226 and as Ullee in 1369, this comes from Ole English *ul leah* and tells of 'the woodland clearing frequented by owls'.

Kent

The name of the county is first seen in 51bc as Cantium. Clearly this is an ancient Celtic name, although the actual meaning is uncertain but has been explained most often as 'the coastal district' or alternatively 'land of the armies'.

Kingsdown

This parish near Deal is listed as Kyngesdoune in 1318. The name comes from Old English *cyning dun* and describes 'the king's hill'.

From Old English *dun geat*, the local name of Dungate describes 'the way to the hill'. Minching Wood was once held by women of God, for the origin of *myncene wudu* tells us this was 'the wood belonging to a nunnery'. Pells Farm features a dialect word, *pell* being used to describe a deeper portion in an otherwise fairly consistent depth. It is easy to see the original Old English *pol* still being used by anglers to describe a 'pool' in a river, which is where fish are more likely to congregate.

Spittal Wood features an element seen in many places, named after the *spital* or 'hospital'. Here that was associated with the church dedicated to St Mary. Stacklands Wood is found as Stokfeld in 1338, quite clearly from Old English *stocc feld* and describing 'the open land where tree stumps abound'.

The Zetland Arms shows a link to the Shetland Islands, probably through a former owner or landlord, this being the old county name for the group.

Kingsnorth Sign

Kingsgate

A name which comes from a landmark event of English history. On the 29th May 1683 Charles II landed at this 'gap in the cliffs', hence the name. This marked the period in our nation's history known as the Restoration of the Monarchy, a day which coincided with the king's birthday and was thereafter known as Oak Apple Day. This name was chosen to mark the famous event when Charles escaped the opposing forces following defeat at the Battle of Worcester two years earlier. He hid in the oak tree at Boscobel in Shropshire while the soldiers walked just feet below him as he hid in the branches with his aide Colonel Carless.

Kingsnorth

Listed as Kingesnade in a document dating from 1226, this is derived from Old English *cyning snad* and describes 'the detached piece of land belonging to the king'.

Ellingham Farm is a local name which is derived from 'the *ham* or homestead of a woman called Aelfwynn'. Westhawk Farm is recorded as West Hauk in 1690, and as West Halke exactly a century later. Here is a name from Old English *west halh* and describing 'the western corner of land'.

Kingston

A common name and one which is still easily seen as 'the royal manor', here the name comes from Old English *cyning tun* and is listed as Cyninges tun in 838 and as Chingestune in 1086.

Black Robin Lane takes its name from the Black Robin public house, this being the soubriquet of a former highwayman who prowled the area. He earned his name from his black attire, which included the colour of his cape, his mask and his horse.

Knowlton

Found in the Domesday record as Chenoltone in 1086, this name is derived from Old English *cnoll tun* and describes 'the farmstead by a hillock'.

CHAPTER ELEVEN – L

Langdon (East & West)

Listings of this name include as Langandune in 861, as Estlanfedoun in 1291, and as Westlangedone in 1291. Here the basic name is derived from Old English *lang dun* and speaks of 'the long hill or down', while the additions are self-explanatory.

Cane Wood is recorded since the sixteenth century, a property of the abbey and thus 'the wood of the canon'. Hollands Hill takes the name of what was originally a road, that telling of 'the lane through a hollow'. Solton is seen as Soltun in 1038 and as Soltune in 1086, this from *sol tun* and referring to 'the farmstead at the muddy pools'.

Leaveland

Found in the Domesday record of 1086 as Levelant, this name was derived from a Saxon personal name and Old English *land* and tells us of 'the cultivated land of a man called Leofa'.

Bagshill is found as Balgameshull in 1313, this from *balg hammes hyll* or 'the rounded or smooth hill by the hemmed-in place'.

Lee

The earliest record is from Domesday when the name appears exactly as it does today. Derived from Old English *leah* and telling of the '(place at) the woodland clearing'.

Shrofield Farm is a local name recorded as Shrafholt in 1240 and derived from *scraef holt* and meaning 'the thicket by a pit'.

Lenham

Listings of this name include Leanaham in 858 and as Lerham in the Domesday record of 1086. This name is derived from Old English *ham* with a Saxon or Jute personal name giving a definition of 'the homestead of a man called Leana'.

Dickley Farm derived its name from Dickley Wood, itself named for the *dic leah* or 'woodland clearing with a ditch' recorded since the thirteenth century. Green Hill speaks for itself, despite being derived from a language first used more than 1,500 years ago and *grene hyll*. Lade Wood comes from *lad*, an Old English term referring to 'a path'. We have reason to believe this was used to bring sheep across this area as there is a thirteenth century reference to Sheplade.

Lewsome Farm speaks of 'the farmstead of a man called Leofric', this common Saxon personal name prefixing *tun*. To the south is Liverton Street, a hamlet named for its beginnings as 'the farmstead of a man called Leoflaed'. Oxley Wood is not recorded before the thirteenth century, making this difficult to define. This solitary early record as Exetheghe suggests an origin of *exen teag* or 'the enclosure for oxen', although this cannot be certain.

Reynold's Wood is derived from former residents, the family of Roger Reynold recorded here in a document dated 1371. Sandway appears as Sandweie around the middle of the thirteenth century, this from Old English *sand weg* and describing 'the way on sandy soil'. Domesday gives the names as Estselve and Westselve, although today east and west have become Old Shelve and New Shelve, both speaking of the settlement on 'shelving land'.

Swadelands is documented as Swallond in 1250, this probably representing 'the agricultural land of the swain or peasant', although *swealwa land* or 'the agricultural land frequented by swallows' should not be ruled out. Timbold Hill has been seen since the ninth century, this coming from Old English *falod ingas* and a Saxon personal name telling us it was 'the fold of the family or followers of a man called Taena'. Torry Hill takes its name from the family of John Terri, recorded here in a thirteenth century document. Waterditch Farm is self-explanatory, coming from Old English *waeter dic* and has hardly changed from the thirteenth century when it appears as Waterdich.

The Red Lion is still the most common pub name in the land. As with the vast majority of coloured animal names the origin is heraldic, the earliest examples would be derived from the coat of arms of John of Gaunt, the most powerful man in England in the fourteenth century, later the reference is to Scotland.

Leysdown on Sea

Seen in a document of around 1100 as Legesdun, this is clearly from Old English *leg dun* and tells of 'the hill with a beacon fire'. The addition of 'on Sea' is very recent, not seen before the Victorian age.

Both Mustards Farm and Mustards Road derive their names from the family of Thomas Smith Musterds, recorded here for the first time in 1679. Newhouse is from Middle English *newe halle* and tells us it was not the original large home here. Shellness Lane could not be better named today for a road which almost skirts the edge of the land as 'the headland where shells abound'.

Pub names include the Seahorse Inn, clearly a maritime reference but not always to the very unfishlike creature known by that name. In early days mariners referred to the walrus as a seahorse and there is a small chance this could have influenced this pub name.

Littlebourne

Found as Littelburne in 696 and as Liteburne in the Domesday record of 1086, this comes from Old English *lytel burna* and describes 'the little place on the river called Burna'. The river is today known as the Little Stour.

Minor names begin with Fishpoolhill Wood, a name featuring three quite different images in the landscape. It has evolved over centuries, beginning in the thirteenth century simply as Fispole. Coming from Old English *fisc pol* 'the pool where fish are kept', the name was later transferred to the hill and, comparatively recently, the nearby wood. Woolton Farm is seen as Woltun in 1197, Old English *wol tun* speaks of 'the pestilence farmstead', probably a place known as a harbour of disease for some time, the most likely being a water source.

Malthouse speaks for itself and was indeed where malt was prepared for brewing. Indeed we even know when this process took place for a document dated 1314 names the resident Robert le Maltmakier. Such a name is a delight for the author for, while we have no notion of Robert's image, we can place the figure in his home and his work and in the right era simply by defining a place name. Found as Holregge in 1220, Oldridge Wood comes from *hol hrycg* and probably refers to 'the ridge bordering a hollow' rather than the literal 'hollow ridge'.

Littlestone-on-Sea

A very recent name created in the nineteenth century. It was not chosen arbitrarily but refers to a coastal feature called Little Stone which was certainly here in 1801, although it has since disappeared under the action of coastal erosion.

The Delf speaks of 'something which is dug' from Old English *delf*, in this case a trench which served as a canal to transport goods.

Of the many visitors here each year most come to ride the Romney, Hythe and Dymchurch narrow gauge railway. Since it opened in 1927 it has carried a regular service throughout the warmer months and even carried a gun and soldiers guarding the coast during the Second World War. This popular line was founded by the man commemorated in the name of the Captain Howey Hotel.

Little Stour (River)

A river which shares its name with the Great Stour, itself discussed under its own entry. This example, also referred to as the Lesser Stour, is found as Stur in 686, Sture in 851, Stura in 1220, and Stoure in 1264. Here is a Celtic or British river name meaning 'the strong one'.

Lydd

Recorded as Jlidum in 774, this name speaks of the '(place of) the gates'. This is from Old English *hlid*, not an actual gate as we would envisage it today but a gap, an entrance or exit point.

Dering Farm is derived from the Dering or Deryng family, recorded as living here in the thirteenth century. Jack's Court is found in the fifteenth century as Jakys Court, this being derived from the family of William Jakes, here in 1348. Nod Wood is first seen in 1569, a rather late record but may show this is from Old English *hnodde* meaning 'lump' and pointing to a feature in the local landscape.

From Old English *rip* meaning 'edge or strip', this place name is first seen as Ripp in a document dated 741. Both East Ripe and West Ripe feature this element, the additions self-explanatory. In the modern name of Wickmaryholm Pit, we have an amalgam of three elements, the last the easiest to define as 'the hollow'. In the remainder is the amalgamation of two names, seen in the sixteenth century as Wigmore and Holme, giving 'the pool at the specialised (dairy) farm' and 'the hollytree' respectively.

The coastal location is reflected in the pub names. Here we find the Ship Inn, and also the Pilot Inn, the latter named for the original use of 'pilot', for the man whose local knowledge was invaluable when bringing vessels into port safely by following the deeper water channels and avoiding sandbanks. The Dolphin is a more obvious link to the sea. Probably named through its reputation as the mariner's friend but also chosen, and for the same reason, as a symbol in a family's coat of arms.

Lydden

Recorded as Hleofaena at the end of the eleventh century, this name comes from Old English *hleo denu* and speaks of 'the sheltered valley' or perhaps we should understand this as a 'valley with a shelter'.

Little London is an example of, what seems to us, the odd sense of humour of our ancestors. This name has existed since at least 1270 and is named ironically to emphasise just how unimportant this place is when compared to just about anywhere else, let alone the nation's capital and one of the largest cities in the world. To the west we find Selstead, from Old English *syle stede* this speaks of 'the miry place'. Stockham is found from the thirteenth century, the name from *stocc ham* and referring to 'the homestead marked by stocks or tree stumps'. West Court Downs and West Court Farm are clearly to the west, with the *court* a French reference to the local manor house. Wickham Bushes comes from two sources, Old English *wicham* refers to 'the dwelling place', with the addition of Middle English *bushoppes* telling us it was 'owned by the bishop'.

Lyminge

Found as Limige in 689 and as Leminges in 1086, this name means 'the district around the River Limen'. Here the Celtic river name, the old name for the East Rother, discussed under its own entry, is suffixed by Old English *ge*.

Eastleigh Court took the name of the *east leah* 'the eastern woodland clearing'. Etchinghill is a name also seen in Goudhurst, however they have significantly different meanings. Seen as Tettingehelde and Tettingheld in the thirteenth century, this is 'the *hielde* or slope of a man called Tetta'. Longage Farm is from Old English *lang hecg*, the earlier record of Longehegge in 1322 clearly showing the origin of 'the long hedge'.

Rhodes Farm comes from Old English *rod*, the name recorded as simply Rode in 1327, speaks of 'the clearing'. We also find Rhodes Minnis, itself seen as Mennessegate in 1327, which represents *gemeanness geat* or 'the way on the land owned communally'. Sibton Wood is a typical Old English name, where the common suffix *tun* follows a Saxon personal name, here telling of 'the farmstead of a man called Sibba'.

Teddars Leas can be traced to 1348, when a document shows this was the *leas* or 'meadow' of the family of Thomas Tyndour.

Lympne

The earliest record of this name dates from as early as the fourth century where it appears as Lemanis. This is a Celtic place name meaning 'the elm wood place' and is related to the name of the River Limen, the old name for the East Rother.

To the west is Court-at-Street, recorded as Cortopstreet in 1530. Here is Middle English *curt hop straet*, telling of 'the road or way to or near the enclosed marshland'. In 1357 John Ffolke was living here with his family, hence the name of Folks Wood. Stone Street is the name given to the paved road running from here to Canterbury, hence this describes 'the stony Roman road'.

Lynsted

Records of this place name include Lindstede in 1212, Lyndestede in 1253, Linstede in 1262, and Lynsted for the first time in 1610. This comes from Old English *lind stede* or 'the place marked by lime trees'.

Locally we find Bumpitt, a name with several early forms including Bomepett, Bonput, Bonipette, Bonypette, and Bompette. These are from Old English *ban pytt*, literally 'the bone pit' and must be a reference to an old grave uncovered here in Saxon times. Claxfield House and Claxfield Farm share an origin in 'the open land of a man called Clac', Old English *feld* suffixes the Saxon or Jute personal name.

Dadmans is recorded as Dodmannys in the sixteenth century and, while no record of this name has survived, it seems to be a manorial name. Lewson Street is a hamlet just off the Roman road and hence the addition of *straet* to 'the stone of the lady'. Its name suggests this was a marker stone, one indicating a boundary on land held by a woman. Loyterton is an old name, one recorded since at least the early fourteenth century and describing 'the *tun* or farmstead of a man called Hlothhere'.

Ludgate describes the *hlid geat* referring to a 'swing gate', something so unusual at the time it was a landmark, hence the place name. Norton is a very common place name, an indication the name has a very simple meaning. This comes from *north tun* and tells of the northern farmstead'. Not seen before 1690, Nouds must be manorial in origin although no records of this name have survived to the present.

CHAPTER TWELVE - M

Manston

A name derived from a Saxon or Jute personal name and Old English *tun* which speaks of this as 'the farmstead of a man called Mann'. The place is documented as Manneston in 1254.

Alland Grange is from *eald land* 'the old agricultural land' which, as the term 'Grange' tells us, was a manor associated with a monastery.

Coldswood is recorded as Coleswude in 1240 and as Colleswode in 1357. Here is Old English *wudu* with a Saxon or Jute personal name and telling of 'the woodland of a man called Col'. Pouces is a local name first seen in the early fourteenth century with the family of Thomas Poucyn.

Margate

Found as Meregate in 1254, this name is derived from Old English *mere geat* and tells us of 'the gate or gap leading to the sea'. Other records give this as St John the Baptist, the dedication of the church, and an alternative name which persisted until the early seventeenth century.

Mutrix Road can be traced back to the thirteenth century, when a man by the name of Alfred Motryk was living here.

Cornhill comes from *corn hyll*, Old English for 'the hill where corn is grown'. Garlinge is a eastern district of Margate, one found as Groenling in 824, and as Grenling in 1254. This is from Old English *grene hlinc* 'the green bank'. Hartsdown was clearly a hunting area for this is from *heorotes dun* 'the hill frequented by harts or stags'. Hengrove comes from Old English *henn graf* 'the grove of trees frequented by hen birds'. Yardhurst is recorded as Yerdhurst in 1240, this describing 'the wooded hill where yards or rods are gathered'.

Northdown is easy to see as 'the northern hill'. Sevenscore features a dialect word as the second element, this refers to 'a pasture or field' and clearly here there were seven such related areas by the time the name was first recorded in the fifteenth century. Westbrook has been seen since the thirteenth century, from *west broc* the name does not mean exactly what it seems for this is 'the western marshland'.

In the Flag and Whistle we have the two iconic items associated with the signalling the departure of a train from the platform. The Spread Eagle is an heraldic image used by many families and even nations, it represented the Roman Empire. Considering the number of important families who adopted the falcon as a part of their arms, it is surprising names such as the Falcon Hotel are not more common.

The Wig and Pen shows this pub was found in the region of the town where the legal profession had their offices. At the Quart in a Pint Pot the message has been said to be a warning not to attempt the impossible, however this does not seem likely as it is hardly an invitation. More likely it simply refers to the quality of the beer – that every pint is as good as two elsewhere. The same cannot be said of the Everybody's Inn for the message is clear, this is the place to be.

At the Lord Byron the sign shows an image of one of our nation's most famous poets, likely as well known for his wild personal life as for his written work. An even more famous writer's name is seen outside the Shakespeare. At one of England's most famous resorts we find the Punch and Judy public house, named after the traditional seaside entertainment for children brought to our shores from Italy in the seventeenth century.

George Brydges has his picture outside the Rodney, the pub using his title of Lord Rodney. This eighteenth century admiral has long list of successes, although his victory over the French in the West Indies at the Battle of the Saintes in 1782 is considered his most important success. While the title is understood in the name of the Princess of Wales public house, ie the wife of the Prince of Wales, there have been many holders of the title and only an image or date can help to identify the one individual which, in the case of Margate, refers to Princess Alexandra of Denmark. Nobody has ever held the title for longer, she became known as such on the occasion of her marriage in 1863 and ended in 1901 when the death of Queen Victoria made her husband king and she his queen.

Medway (River)

The earliest surviving record may date from the eighth century where it appears as Medeuuaege, however the name is from a much earlier period. This seems to be a combination of the Celtic river name Wey, a name of obscure etymology, and Celtic or Old English *medu* meaning 'mead' and a description of the water, either its colour or sweetness.

Unlike place names, almost always a description of the local area, rivers and streams are not isolated in one particular area. For example there are three rivers called the Derwent in the land, the name describing the oak trees along the banks. With almost two hundred miles of banks there can never have been oak trees growing along every mile for the landscape simply could not sustain these massive trees.

All rivers age, a young vibrant and fast flowing stream, slows and meanders through lowlands, before emptying into a larger river or the sea where it may be difficult to see any movement whatsoever. When river names reflect the nature of the river - in its simplest form the 'white river' is the young turbulent bubbling stream, while the 'dark one' is the older version with silted or muddy depths - it is clear the same name cannot be applied to the same river. However few rivers have more than one name today, while historically, when people would be unlikely to see more than the one sizable river in their entire lives, finding the name of the river depended upon where and of whom the question was asked. One tributary of the Medway is a good example for alternative names only fell out of favour quite recently.

The modern River Bourne comes from *burna* a highly simplistic name meaning simply 'stream'. Yet as recently as the eighteenth century no less than four alternative names, form two different origins, are documented: the Busty or Buster refers to the tendency of the river to flood, while the Sheet or Shode comes from *shode* 'the branch of a river' and thus describing it as a tributary of the Medway. Here we see three origins describing three sections of the Bourne

– *burna* when it was young and a mere stream; Buster in middle age when it crosses a floodplain; and *shode* at its confluence with the Medway.

The Bradbourne is another tributary, one which is easily seen as 'the broad stream'.

Mersham

Records of this name include Mersaham in 858 and as Merseham in 1086. This features a Saxon or Jute personal name and Old English *ham* and describes 'the homestead of a man called Maersa'.

Bilham Farm began life as 'the hemmed-in land shaped like a sword', from *bill hamm*. Kingsford Street tells of this as 'the open land of the king'.

The Farriers Arms is a reminder of the days when horsepower meant simply that and a pub was invariably next to the village blacksmith, the name advertising such.

Milstead

Found as Milstede in the late eleventh century, this name seems to come from Old English *middel stede* and refer to 'the middle place'.

Hogshaw Wood is a minor name from Old English *hogg scaga* telling of 'the copse frequented by hogs'.

Minster

One of two places in Kent of this name, this example being near Ramsgate. This name is derived from Old English *mynster* and referring to 'the monastery or large church'. This place is found as early as 694, where the name is recorded as Menstre.

Cottington Hill is seen as Cotmanton in 1278, describing 'the farmstead of the cottagers' and shows both the hill and Cottington Road, which runs through here, were named from the original minor place name. Sharden's Farm points to this as 'the woodland pasture where dung was collected', the name derived from *scearn denn* and first recorded in the twelfth century.

Minster

The second example of this place name in the county, this example being found near Sheerness. As with the previous entry this comes from Old English *mynster* and describes 'the monastery or large church'. Furthermore the earliest record of this name is also exactly the same as the previous entry, although here the record of Menstre does not appear until much later, in 1203.

The local name of Durlock is from Old English *deor loca* or 'the enclosure for animals'. Scarborough Drive speaks of its position as 'the stronghold at the cleft or gap' from Old English *sceard burh*. In Domesday the name appears as Tepindene and today as Tiffenden, both speaking of this as 'the *denn* or woodland pasture of a man called Tippa'. Watchester Farm is derived from Old English *aet thaem waetere* meaning 'at the water', that it evolved with the suffix -chester is down to other place names in the area associated with the Romans.

In the Prince of Waterloo is a pub name remembering Gebhard Leberecht von Blucher, Prince of Wahlstadt (1742-1819), who commanded the Prussian Army at Waterloo.

Mongeham (Great & Little)

A name recorded as Mundelingeham in 761 and as Mundingeham in the Domesday record of 1086. Here a Saxon or Jute personal name either combines with Old English *ing ham* speaks of 'the homestead associated with a man called Mundel', or the Old English is *ingas ham* and describes 'the homestead of the family of followers of Mundel'.

Maydensole Farm is a local name which first appears in 1676 as Maidensole. With few records available it is difficult to be certain but *maegdene sol* meaning 'the muddy pool of the maidens' seems the most likely. Pixwell is not seen before 1799, where it appears as Pigsole. This from Middle English *pigge sole* telling of 'where pigs wallow'.

The pub named the Leather Bottle is a pointer that glass is a comparatively recent material when producing bottles for drink as they were expensive and broke quite easily. Making them of leather and sealed with pitch or tar, they could withstand quite rough treatment and lasted a considerable time.

Monkton

A name from Old English *munuc tun* which speaks of this as 'the farmstead of the monks'. The name is recorded as Munccetun in 960 and as Monocstune in 1086.

Plumstone Farm is found as Plumstede in 969 and as Plvmestede in 1086, these from Old English *plume stede* and telling of 'the place of plum trees'.

CHAPTER THIRTEEN – N

Nackington

Listed as Natyngdun in the late tenth century and as Latintone in Domesday, this name comes from Old English *naet ing dun* and describes 'the hill at the wet place'.

Local names begin with Heppington House, this found as Hebbinton in 1181, Hebynton in 1346, and as Hepyngton in 1407. These early forms show this is from 'the farmstead associated with a man called Hebba'. Sextries Farm occupies a site once owned by the monks of St Augustine's in Canterbury, the name indicating it was once home to a sacristy – a room where items are kept until wanted for church ceremonies, including clothing.

Newchurch

A name from Old English *niwe cirice*, this name describes 'the new church' and is listed as Neverce in Domesday in 1086.

Newenden

For once the Domesday record is little different to the modern form, this name is found in the great tome as Newedene. Derived from Old English *niwe denn* and refers to 'the new woodland pasture'.

Copthall is recorded as the modern form as early as 1702, this being from Middle English and describing 'the capped hall', a thought to be a reference to a particularly high roof. Listed as Frogherst in 1338, Frogs Hill seems to have been derived from *frogga hyrst*, an Old English name describing 'the wooded hill where frogs are found'. However it must be said this does seem an unlikely environment to find a large frog population, hence this probably represents a lost place name later transferred here.

Lossenham is found as Hlossanham in a document dated 724, this being from Old English *ham* with a Saxon personal name and referring to 'the homestead of a man called Hlossa'.

Newington

Two parishes of this name in Kent, this example being near Hythe. Listed as Neventone in 1086, the Domesday record showing this to be from Old English *niwe tun* and describing 'the new farmstead'.

Asholt Wood has changed little since the Old English *aesc holt* 'the thicket of ash trees'. Meaning 'the newly cultivated land' from Old English *braec* comes the modern name of Breach. The family of Brockman are recorded here during the seventeenth century, giving their name to Brockman's Bushes. Longport Farm is from Old English *lang port*, not quite what it would seem today but speaking of 'the long market place'. Wormdale comes from Old English *wyrm dael* and tells of 'the valley frequented by snakes', first recorded as Wurmedele in 1185.

Newington

The second example of this name, this example being near Sittingbourne. Listed as Newetone in Domesday in 1086, this comes from Old English *niwe tun* and literally refers to 'the new farmstead' but should probably be understood as 'the newer farmstead'.

Two local names explain themselves, for Nunfield House and Nunfield Farm occupy land once held by a priory of Benedictine nuns. In the thirteenth century William de Frogenhall was granted land now marked on the map as Thrognall. His surname comes from Frognal, a hamlet in the parish of Teynham and defined under that listing.

Newnham

Documented as Newenham in 1177, this is derived from Old English *niwe ham* and describes 'the new or newer homestead'.

Champion Wood takes the name of the lords of the manor, the de Campania family were here at the end of the twelfth century. Smalldane Wood was named for its location in the *smael denu* or 'small (or narrow) valley'. Sprats Hill takes its name from the family living here in 1231, the head of the household one Godefridus Sprot. Stuppington comes from Old English *stupes dun* or 'the hill with a steep slope' and seen as Stupesdone in 1253.

Noah's Ark

What might seem an odd or even unique place name has two other examples found in Cheshire and Derbyshire. We can trace the Kent version to a house built here around the beginning of the eighteenth century, the name first seen on an Ordnance Survey map from 1819, however this does not explain the origin of the name. Clearly the one thing associated with the name is a variety of animals and it seems this must be why the name was chosen, for some unknown connection with animals.

Nonington

Recorded as Nunningitun at the end of the eleventh century, this name comes from a Saxon or Jute personal name and Old English *ing tun*. Together this refers to 'the farmstead associated with a man called Nunna'.

Locally we find Curlswood Park Farm, seen as Crudeswod in the middle of the thirteenth century and featuring a personal name and Old English *wudu* telling us it was 'the wood associated with a man called Crud'. From the Old French *freide ville*, Fredville literally means 'the cold place', this is still an exposed position. Found as Frogenham in 1270 and as Frogeham in 1346, the modern name of Frogham is from *froggena hamm* here understood as 'the water meadow frequented by frogs'.

Holt Street comes from Old English *holt* and refers to 'the thicket'. Kittington Farm is recorded from the thirteenth century, an Old English name from *cyte ham tun* and telling of 'the cottage home farm'. St Alban's Court reminds us this was land granted to the Abbot of St Alban's during the twelfth century.

Northbourne

Listed as Nortburne in 618 and as Norborne in 1086, this comes from Old English *north burna* and simply refers to 'the northern stream'.

Haddling Wood is a place name first recorded in 1240 as Hedelinge. Here a Saxon or Jute personal name is suffixed by Old English *ingas* and refers to the '(place of) the people of Haeddel'. Note this refers to the people, not the man himself. Thus the named individual was not here, indeed it was probably named posthumously. From Old English *maed aecer* comes 'the meadow land' and seen today in the name of Minacre Farm, undoubtedly the pronunciation and therefore spelling influenced by nearby Lenacre.

Old English *stan aern* or 'the stone house' has today become Stoneheap Farm, likely this shows there was a rebuilding at one time with some stones being left over.

CHAPTER FOURTEEN – O

Oare

A name recorded in Domesday as Ore in 1086, this name is derived from Old English *ora* and describes the '(place at) the shore or slope'.

Ospringe

Found as Ospringes in Domesday, this name comes from Old English *or spring* and referring to the '(place of) the spring'.

Little Coxett is recorded as Cocsete in 1247 and as Coksete in 1466, a name from Old English *set* with a Saxon or Jute personal name and telling of 'the dwelling (literally seat) of a man called Cocca'. Meaning 'the cultivated land of a woman called Aelfwaru', Elverland is recorded as Elverlond in 1247 and as Elveyrlond in 1346. A document dated 1380 has Hail Beech Wood listed as Caldhelbeche and also as Ealdhebeche. While these only have a different initial letter, it does make for two different meanings. The former is from *cald hyll bece* 'the beech trees of the cold hill', while the latter is from *eald hyll bece* 'the beech trees of the old hill'. Whitehill is from Old English *hwit hielde* and describes 'the white slope', the soil being predominantly chalk.

Jarvis Down Wood is a place associated with the family of John Gerueys in 1327. Daniel Judde built a house in 1652, that house known as Judd House and the name transferred to the present Judd's Hill. Kennaways remembers the family of Kenewy, here by the thirteenth century. In Painters Forstall we have a name seen as home to Robert le Peyntour in 1278, later the addition of a dialect word to show it was located 'in front of farm buildings'. Plumford Farm can be traced back to Old English *plume worth* telling of 'the enclosure marked by plum trees'.

Porter's Lane remembers the family of Hugh le Porter, here by 1270 and well documented in the area around this time. Putt Wood is found as Putewude in 1164, this featuring the suffix *wudu* and telling of 'the wood of a man called Putta'. Queen Court was once the property of Anne of Bohemia, the queen and consort of Richard II and first named as such in a document of 1405. Rushett Lane was seen for the first time in 1782, the name is from Middle English *rishette* or 'rushy place'. In the case of Tassell's Wood, this takes its name from the family of Thomas Tassell, recorded here by 1522. Willow Wood means almost what it seems, there was *wilig* or 'willow' growing here but could hardly be described as a wood in the traditional sense hence the modern deciduous woodland took on the earlier name.

Otterden

A name derived from a Saxon or Jute personal name and Old English *inga denn*, this refers to 'the woodland pasture of the family or followers of a man called Oter'. This name appears in Domesday as Otringedene.

Bunce's Court is a lasting reminder of former hall owner John Bunce, here in 1503, although today it is often known simply as The Hall. From Old English *fyrhthe* or 'woodland' comes the name of Frith. Limekiln Wood speaks for itself as the woodland near 'the lime kiln'.

CHAPTER FIFTEEN – P

Paddlesworth

Listed as Peadleswurthe in the eleventh century, this is derived from a Saxon or Jute personal name and Old English *worth* and describes 'the enclosure of a man called Paeddel'.

The hamlet of Arpinge to the south seems to represent a Saxon personal name with Old English *ing* and telling of the '(place) associated with the people of Eorpa'. However it is possible this is not a personal name but represents Old English *eorpingas* and describes the '(place) associated with the dusky or swarthy ones'. By 1348 John Jordan and his family were recorded at what is now known as Jordan's Wood.

Redsole Farm is derived from Old English *hreod sol* or 'the muddy pool where reeds grow'.

Patrixborne

Listed simply as Borne in Domesday, this comes from Old English *burna* and speaks of 'the stream', a reference to the Little Stour. Later we find the record of Paricburn in 1215, a manorial addition showing this place was held by one William Patricius from the twelfth century.

Pegwell Bay

A name not seen before the document dated 1799, a late record which is not easy to define. The most obvious origin is Middle English *pigge well* and literally 'the pig spring'.

Petham

Listed as Piteham in the Domesday record of 1086, this name comes from Old English *pytt hamm* and describes 'the hemmed in land with a hollow'.

Ansdore is recorded as Aginsdore in 1240, the name telling of 'the door or pass of a man called Aegene'. Richdore shares the suffix but was associated with one Hugh le Riche, here in 1240 but not seen as part of the place name until after the sixteenth century. Middle English *grene* in Broadwaygreen Farm can be ignored, this is a late addition to *brad weg* 'the broad way'. Buckholt Farm is derived from *boc holt*, the Old English for 'the thicket or by the beech trees'. Debden Court is recorded as Depedene in 1403, from Old English *deop dene* this is 'the deep valley'.

Gogway is alongside an old route, as the name says this is 'the *weg* or way of the *gog*'. A *gog* being a dialect word for a 'bog or mire'. Grandacre Farm is a combination of Old French *grand* and Middle English *aecer*, together these speak of 'the large cultivated area of land'. Hault Farm was recorded as Haute in 1278, taking the name of the family of Ivo de Haut who were here by 1270. Waltham is derived from *weald ham*, 'the homestead in the now cleared forest' recorded as Wealtham in a document dating from the eleventh century.

Hobday's Wood is a family name, the Hobday family owning nearby Anvil Green and associated with this area in some way. From Old English *cyninga* feld comes 'the open land of the king', today known as Kenfield Hall. Sappington Court originated from *saeping dun* or 'the hill where saplings grow'. Sheep Court was recorded as Shytecourt in 1346, this from Old English *sceat* and referring to 'a nook or corner of land'. That the name has evolved to 'sheep' is both late and seemingly deliberate.

Swarling has been recorded since the tenth century, the name coming from *sweordlingas* and telling of 'the settlement of the swordsmen'. Wootton Farm features a common minor place name, this from *wudu tun* or 'the farmstead at the wood'.

Pilgrims Way

Not only one of the most famous routes in Kent but also in the country. It is named as it is the traditional route followed by those journeying to the tomb of St Thomas a Becket in Canterbury Cathedral. It runs along the line of the North Downs between the border with neighbouring Surrey to Felixstowe, running both ways as pilgrims travel to Canterbury from both directions. However the route was in use well before the reign of Henry II and Becket, his Archbishop of Canterbury, and was likely a trading route in use well before the arrival of the Romans.

Pluckley

Often cited as the most haunted village in England, the place name comes from Old English *leah* following a Saxon or Jute personal name. Here is 'the woodland clearing of a man called Plucca'.

To the east Dowle Street gets its name from the family who lived there between the fourteenth and sixteenth centuries, their surname recorded as Doul, Dowele, or Dowle depending upon which document is consulted. Sheerland is from *scir land*, that is 'the agricultural land belonging to the shire' and not privately owned.

Pluckley has a well-deserved reputation as the most haunted place in England. At the Dering Arms the pub has its own White Lady, whose appearances are so clear she is often mistaken for a customer. The pub is named after the family who have a long history here. The Tudor Lady was a member of the Dering family, she committed suicide by eating poisonous berries and is said to walk the gardens calling to her dogs. Other members of the family haunt St Nicholas' church, known as the Red Lady and the White Lady.

Black Horse public house at Pluckley and ......

..... the sign hanging outside.

St Nicholas Church in Pluckley

Postling

Found in the Domesday survey as Postinges and as Postlinges in 1212, this comes from a Saxon or Jute personal name and Old English *ingas* and speaks of the '(settlement of) the family or followers of a man called Posell'.

Preston

One of the most common of English place names, from Old English *preost tun* this is 'the farmstead of the priests'. The name is recorded as Preostatun in 946, Prestetone in the Domesday record of 1086, and Prestone in 1253.

Locally we find Deerson Farm, a name recorded as Deyredeston in 1290 and derived from a Jute or Saxon personal name with the common Old English *tun* and giving 'the farmstead of a man called Daegred'. Although not seen before 1538, Hardacre Farm is a name dating from long before the Conquest when the Saxons referred to it as *heard aecer*, still self-explanatory as 'the land which is difficult to work'.

Hoaden is easily seen as *heath dun* 'the hill covered by heathland'. Ladydown Farm is literally 'lady hill' and tells us this farmland was once the property of a woman. Perry Court is a name first seen in the Domesday record of 1086. Here the listing as Perie is from Old English pyrige or 'pear tree'. Luckett Farm was seen as Luuecote in the early thirteenth century, showing this to be from 'the cottage of a man called Lufa'.

Santon Farm is listed as Sywent in 1154, this from a Saxon personal name and the most common suffix *tun* and giving 'the farmstead of a man called Sigewynn'. Ware comes from an element *wer*, a common part of a name and occasionally seen on its own. Here 'the weir' refers to that constructed on the Wingham River, almost certainly to feed a mill race. West Marsh speaks for itself as 'the western marshland'.

CHAPTER SIXTEEN – Q

Queenborough

Listed as Queneburgh in 1376, this name comes from Old English *cwen burh* and describes 'the borough named after Queen Philippa', she was the consort of Edward III.

Elmley to the south comes from Old English *elm leah*, listed as Elmele in 1226 and Elmeley in 1270 this is 'the woodland clearing by the elm trees'. Rushenden to the south is named for being 'the hill covered with rushes' and derived from *riscen dun*. This 'hill' is not easily spotted for the region is quite flat, as indeed is indicated by the suggestion it is covered in rushes which would require a consistently wet soil. Hence the 'hill' here is simply raised ground. Stray Marshes is first seen in a document dated 1553, which seems to be from *stregan* and used to describe the 'spreading' of the water as it formed the watery border between this and the Hoo Hundred.

The Flying Dutchman is named after the legendary ghost ship said to foretell imminent doom for sailors spotting it off the Cape of Good Hope. The Queen Phillipa is named after the queen consort of Edward III and mother to fourteen children, including the Black Prince and the most powerful man in England during the fourteenth century, John of Gaunt. One of the most famous battles in history was the inspiration when naming the Trafalgar Hotel.

CHAPTER SEVENTEEN – R

Ramsgate

Recorded as Remmesgate in a document from 1275, this is derived from Old English *hraefn geat* and speaks of 'the gap of the raven', the gap is of course a natural cleft in the cliffs. However the first element may be a personal name, for Hraefn was a popular name for both Germanic and Scandinavian cultures.

Cliffsend is self-explanatory, the name recorded since the early thirteenth century. Ebbsfleet is derived from the Old English for 'the *fleot* or tidal inlet where hips grow'. Hereson is seen as Heyredeston in the early thirteenth century, this being 'the farmstead of a man called Heahraed'. Sprattling Street recalls former resident Adam Spratlyng, here with his family in 1292.

The Queen Charlotte public house remembers the consort of George III. Her marriage was not a particularly happy one, especially in the early years, although they did manage to produce 15 children, all but two surviving to adulthood, a tremendous achievement for the time. She was a popular royal, patron of the arts, and a greatly respected amateur botanist. In the East Kent Arms we find a link to the East Kent Railway, which once served this area.

The Cherry Orchard can only have been named from just that, although whether it stood here or was owned by the landlord is not clear. Charles Pratt, 1st Earl of Camden gave his name to the Camden Arms. This eighteenth century politician became lord chancellor and held the land which was developed as Camden Town. The Little Brown Jug was named because of the image of the container and the brown liquid it contained, although the popularity of the traditional ditty of the same name was undoubtedly a consideration.

In the Montefiore Arms we find a name remembering Sir Moses Montefiore, a Jewish banker, financier, philanthropist and former Sheriff of London who gave vast sums of money to promote industry, education and health. Renowned for his sharp wit, it is said he was once seated next to a vociferous anti-Semite at a dinner party who told him he had just returned from Japan where they had neither "pigs nor Jews". Montefiore offered to return with him so Japan could boast an example of each. He died in Ramsgate, aged 100.

At the Blazing Donkey we find an old term which simply refers to a 'braying donkey'. In the early seventeenth century a royal visitor came to visit the newly-crowned James I of England and VI of Scotland, his brother-in-law Christian IV giving his name to the King of Denmark. Imagination continues to provide pub names and in the case of the likely unique Hundred and Eighty invites patrons to a game of darts.

Admiral Sir John Jervis defeated a large Spanish fleet at Cape St Vincent in 1797. Later he was made First Lord of the Admiralty and oversaw the rapid advancement of arguably our greatest hero, Horatio Nelson. He is remembered in the pub named from his title, the Earl St Vincent. Another maritime link is seen in the Foy Boat, these being a general term applied to the small boats which went to sea to help vessels in need of help, not in the same way as a lifeboat but ferrying supplies and running errands.

Reculver

Found as Regulbium around 425 and as Roculf in 1086, this is a name from the ancient Celtic tongue meaning 'great headland'.

Lovestreet Farm is a local name derived from the family named Love, here before 1327. The King Ethelbert is named after the King of Kent in the late sixth century, a man who was converted to Christianity by St Augustine and later the two men worked together to make Canterbury the religious centre it has been for the next fourteen centuries.

Richborough

A document from 1197 lists this name as Ratteburg. This may seem old at eight centuries, yet the previous record is over a millenium older and dates from Rutupiae around ad150. This name is a combination of an old Celtic term thought to mean 'muddy waters', with Old English *burh* or 'stronghold'.

The local name of Rubery Drove is seen as Ruweberg in a document dated 1240 and as Rewbergh in 1382. These are derived from Old English *ruh beorg* and describe 'the rough mound or hill'.

Ringwould

From a Saxon or Jute personal name and Old English *inga weald*, this name speaks of 'the woodland of the family or followers of a man called Redel or Hrethel'. The name is found as Roedligwealda in 861.

Local names include Martin, from *mere tun* describing 'the farmstead by a pool'.

Ripple

Listed as Ryple in 1087, this comes from Old English *ripel* and describes the '(place at) the strip of land'.

To the southeast, rather than the south, comes Sutton or 'the southern farmstead'. Winklandoaks Farm acquired its 'oaks' element comparatively late and requires no explanation. Recorded as Winekelande in 1232, this referring to 'the agricultural land of a man called Wineca'.

Rodmersham

Recorded as Rodmaeresham around the end of the eleventh century. Here is a Saxon or Jute personal name with Old English *ham* and describing 'the homestead of a man called Hrothmaer'.

Cheney Wood is listed as Chayney Courte in 1549, the name from the family of Robert Cheyne who were here by 1489.

Pitstock Farm is derived from Old English *pise stoc*, 'the place where peas grow' being recorded as Pistok in 1254.

Rolvenden

Telling us of 'the woodland pasture associated with a man called Hrothwulf', this is derived from a Saxon or Jute personal name and Old English *ing denn*. Domesday records this name as Rovindene in 1086 and as Ruluinden in 1185. Nearby Rolvenden Layne has nothing to do with road or ways, this refers to an area to the southeast which is arable land today but will have, as the name suggests, 'lain fallow' at times.

Records of Chessingdenn, Chestenden, and Chesynden in the thirteenth century lead to the origin of Chessenden from 'the *denn* or woodland pasture of a man called Ceasta'. Corn Hill is a corruption of Old English *cran wielle* or 'the spring frequented by cranes', a name with a quite different meaning than that suggested by the modern form. With the earliest surviving record as Farnehille from 1440, Farnell Wood comes from *fearn hyll* and easily seen as 'the hill where ferns grow'.

Freezingham is a reminder of 'the settlement of the Fresingas', a tribe descended from the Frisians on the European mainland. Gibbon's Wood is named after the family of Robert Gybbon, they are documented as living here in 1509. Maytham is recorded as Maiham in 1185 and as Meyhamme in 1242, from *maegtha hamm* and speaks of 'the hemmed in land where mayweed grows'. Sheaf Wood takes its name from the family of Richard Sheafe, here before 1647. Winton Farm features a Saxon personal name and the suffix *tun* and tells of 'the farmstead of a man called Wighelm' and seen as Wilmenton in 1334 and Wylmynton four years later.

When it comes to pub names we begin with the Star, originally a religious reference from the best known Bible story of them all. However today the star of Bethlehem is not the main reason for the name today, although doubtless it was the reason it was chosen for the main device in the arms of the Worshipful Company of Innkeepers.

Romney (Old & New)

Listed as Rumena in 895 and as Romenel in 1086, this place name was originally a river name. From Old English *ea* 'river' this suffix appears to follow the element Rumen, itself an old name for Romney Marsh meaning 'the broad one'.

At New Romney is Bow Bridge, originally from *boga* and found as Boghe in 1348, does describe 'a bow' although today we would refer to the bridge as being 'arched'. Bybrook is not as simple as it seems, this is from *beo broc* or 'the marshy ground with many bees', almost certainly attracted here by the numerous marsh flowers. Welland Marsh is a name first seen in the thirteenth century, this from Old English *weall land* or 'the agricultural land near a sea wall'. Yoakes Court comes from Old English *geoc* and refers to 'the yoke of land', an early measurement of area.

Cockreed Lane describes 'the cleared land of a man called Cocca', the lane would have lay alongside this area. Honeychild comes from Old English *hunig celde* and describes 'the honey spring', a reference to its sweetness. Rosecroft remembers the family of John Rose, here before 1374. Wallingham is found as Walengeham in 1397, the name showing this had formerly been a British settlement and telling of 'the homestead associated with the place of the Britons'. Old English *wealh* literally means 'foreigner' but evolved to become the English name for the Welsh.

The Cinque Port Arms is a pub name reminding us this was one of the five ports nominated by Edward the Confessor to provide ships and men to defend the nation in times when our island was under threat of attack. Along with Romney, Hythe, Sandwich, Hastings and Dover were the five, although only the last two are still ports for the others silted up long ago. Another maritime name is easily recognised in the name of the Sea Horse. Today the image is that of the fish probably best known for looking very unfishlike. However as a pub name it first referred to a fabulous beast which had the front part of a horse and the hind parts of a fish or marine mammal. It was also used for a walrus, which was probably the original beast when seen in the water.

Rother (River)

This watercourse is an example of back-formation, the place being Rotherfield in Sussex and the source of the river. The earlier name comes from a British word, *lem* referring to 'the elm tree'. The name is recorded as Liminel in 1180, Lymmene in 1279, Lymme in 1474 and first seen as Rother in 1575.

Royal Military Canal

A modern name for a modern water course. Running from Hythe to Rye, it was dug in the nineteenth century as part of the defences along the south coast to help repel the threat of invasion by the French under Napoleon.

Ruckinge

Listed as Hroching in 786 and as Rochinges in 1086, this name probably comes from Old English *hroc ing* and referring to 'the place frequented by rooks'. However it should be noted that Hroc is also a personal name and thus 'the place associated with a man called Hroc' cannot be discounted.

Local names include Freeland Wood, from Old English *freo land* this is easily seen as 'the free land', that is having no payable rent. Lodgeland does indeed speak for itself as 'the farmland of or near the estate lodge'. However it was only known as such from the eighteenth century and until at least 1790 was Cobbes Place, after John Cobbe, a man first recorded here in a document dated 1313.

From Old English *ruh stocc* comes Roughstocks Wood, it describing the place of 'the rough stocks or stumps'. Warehorne can be traced to a document dated 830, where Werahorna is seen as from Old English *wer horn* and speaks of 'the spur of land near a weir'. Hamstreet comes from *ham* 'the road at the homestead'.

Hamstreet village sign

Hamstreet's Dukes Head…..   *….. and it's sign.*

CHAPTER EIGHTEEN – S

St Lawrence

A name listed as Church of Sanctus Laurencius in 1253, this name comes from the dedication of the local church.

Bethlehem Farm is named for it once being a possession of Bethlem or Bedlam Hospital in London. Found as Osingehull in 1240 telling of 'the hill of the family or followers of a man called Osa', where the Saxon personal name is followed by Old English *inga hyll*.

St Margaret's at Cliffe

Recorded as Sancta Margarita in Domesday and as Cliue at the end of the eleventh century, todays form unites both names in 'the church dedicated to St Margaret at the place known as Cliffe'. This place name is derived from Old English *clif* and, at least here, certainly describes 'the cliff'.

Bere Farm is listed as La Bere in 1235 and as Bere in 1270, derived from *byre* it reminds us this was named after 'the shed or hovel'. Reach Court comes from *hrycg* and describing 'the ridge'. In the case of Wanstone Farm the common *tun* follows a personal name and, with this documented as Wansieston in 1233, tells of 'the farmstead of a man called Wensige'.

St Mary's Bay

Documented as Seyntemariecherche in 1240, this name is taken from St Mary in the Marsh. Here the latter name takes the dedication of the church and adds 'marsh' from Old English *mersce*.

A little inland we find Brodnyx, a name derived from the family of Richard Broadnex who were here by 1467.

Here is a pub named the Bailiff's Sergeant. It was dedicated to the Jurates of the Level of Romney Marsh and the Office of the Bailiff's Sergeant by the brewers, recognising the work they undertook in draining the marshes and building dams to hold back the tides of the sea.

St Peter's

Listed as Borgha sancti Petri in 1254, this 'borough of St Peter' is named after the dedication of its church.

The local name of Calais Wood can only have been brought here as a family name of someone heralding from the best known French port.

Stone House and Stone Farm are both indicators there is a greater than normal amount of stone in the local soils.

Once there were more than 600 examples of pubs called the Red Lion in this country. While there are fewer today, it remains one of the most common pub names. As with many oddly coloured animals the origin is heraldic. It began as representative of John o' Gaunt, the most powerful man in England in the fourteenth century. He sired children who gave rise to a line who came to rule Scotland and, at the accession of James VI of Scotland as James I of England, were rulers of England too. As a pub name it is probably used to represent Scotland, however the ultimate origin is from John o' Gaunt.

Saltwood

Found as Sealwuda at the end of the tenth century and as Salteode in the Domeday record, this name is derived from Old English *sealt wudu* and tells us this was 'the wood where salt was made or stored'.

The name of Brockhill Park has changed little since the thirteenth century, this coming from *broc hyll* 'the hill near marshy ground'. Heane Barn and Heane Wood may not be of great importance today, yet in the Saxon era it was the meeting place of the Heane Hundred. The name also alludes to the meeting place for *hiwun* means 'family, household' and used in the sense 'community'. From a Saxon personal name with Old English *ingas*, Pedlinge describes the '(place) associated with the people of a man called Pydel'.

Thorn Wood appears as simply Thorne in 1278, this a clear reference to 'the thorn bushes'.

Sandgate

From Old English *sand geat* and telling us of 'the gate or gap leading to the sandy shore', this name was recorded exactly as it appears today as early as 1256.

The minor name of Enbrook is seen as Einesbroc in 1166 and Eynesbroc in 1242, this is 'the brook of a man called Aegen'.

Sandhurst

With the earliest record of this name appearing as Sandhyrste around the end of the eleventh century, it is easy to see this as coming from Old English *sand hyrst* and describing 'the sandy wooded hill'.

Alderden Manor gets its name from 'the woodland pasture of a man called Ealdhere'. From Old English *box hyrst* comes Boxhurst or 'the wooded hill of box trees'. Found as Cellyndenn in 1292 and as Chillindene in 1327, Challenden comes from a Saxon or Jute personal name and *denn* and referring to 'the woodland pasture of a man called Ceolla'.

Chandler's Barn derives its name from the family of John Chaundeler, recorded here in 1431. Cowbeech Farm is first recorded as Cobeech Green in 1690, a name from Middle English *cou beche* 'the cow beech' and the spreading branches of a beech tree under which cows could be seen sheltering. Ethnam has changed since the early record of Echenham in 1313, this being from *acen hamm* and describing 'the hemmed-in land where oaks grow'.

Herndon is recorded from the early thirteenth century, all forms showing this to be from *hyrne denn* and 'the woodland pasture in a corner of land'.

Hexden is found from the thirteenth century, the name also transferred to nearby Hexden Channel, speaking of 'the *denn* or woodland pasture of a man called Haecci'. Holman's Wood is a reminder of the family named Holman, in the neighbouring parish of Newenden in the sixteenth century.

Lamberden has been recorded since the late thirteenth century, a name from *lam denn* or 'the loamy woodland pasture'. Linkhill has been recorded since the early thirteenth century, a name from *hlinc denn* and describing 'the pasture on sloping ground'. Oakenden is listed as Akynden in 1278, from *acen denn* and telling of 'the woodland pasture marked by oak trees'. Silverden Lane is recorded from the thirteenth century, the name describing 'the woodland pasture of a man called Selebeorht'.

Trillinghurst is found as Therlingehurst in 1254 and Terlingeherst in 1334, this from Old English *hyrst* with a Saxon tribal name telling of 'the wooded hill of the trillers of songsters'.

Sandwich

Seen as Sandwicae in the early eighth century and as Sandwice in the Domesday record, this name comes from Old English *sand wic* and refers to 'the sandy harbour or trading centre'.

Sandwich was one of the original Cinque Ports, a term coming from Old French *cink porz* meaning 'the five ports'. Originally found as De Quinque Portibus in 1191 the modern form appearing for the first time towards the end of the seventeenth century, the other four were Hastings, Dover, Romney, and Hythe.

To the east Ash, recorded as Ece in Domesday, comes from *aesc* and tells us there was once 'an ash tree' which would have been very prominent in the landscape. Boatman's Hill can be traced back to 1526 when Henry Bateman's family were resident here, the modern spelling is from an association with the port. Marshborough is recorded in Domesday as Masseberge in 1086, from Old English *beorg* and a Saxon personal name describing 'the tumulus of a man called Maessa'. Great Poulders and Little Poulders share a name from Old English *polra* telling us this was 'water-logged land'. Stonar appears as Stanoure in 1225 and as Store in 1227, this from *stan ora* or 'the stony bank of land'.

Pubs include the Red Cow, for once not a coloured animal from an heraldic source but a reference to the brown cattle whose colouring is certainly a reddish hue.

Sarre

Listed as Serrae in a document date 761, this name has never really been understood but could prove to be an early Celtic name and related to Latin *serare* in meaning 'to close or shut'. In this case it would describe dams erected on the stream.

The local name of Marshside would appear to add to the watery theme, although it may have been brought here as a surname by the family of Wilmari de Marisco, here in 1225. St Nicholas at Wade clearly took the name of the dedication of the church and added *waed*, 'a ford or wading place'. Wade Marsh and Wade Marsh Stream show just what that 'wading place' crossed.

Seasalter

Recorded in Domesday as Seseltre, this tells us of 'the salt works on or by the sea' and is derived from Old English *sae sealt aern*.

There are many pubs named the Blue Anchor in the country. While nearly every example is close to the coast, it is not a maritime reference but a religious one. The colour represents Christianity and the anchor hope, the latter taking a line from Hebrews 6:19 where St Paul speaks of one's faith being an anchor through life's storms.

Sellindge

Found in the Domesday book as Sedlinges, this name is probably derived from Old English *sedl ingas* and telling of 'those sharing a building'.

Alternatively this could represent *sedl ing* and then it would be understood as 'the place at the building'.

Carpenters Wood is named from the family of Elias le Carpenter, first recorded here in 1313. While Harringe Court is certainly from Old English, it is difficult to know if this should be understood as simply the '(place of) the family or follower of a man called Here', if the first element is a personal name, or if from *here* 'the place of the tribe of fighting men'. Recorded since the thirteenth century 'the moorland place' comes from *mor stoc* and is seen today as Moorstock.

Somerfield Court was built on 'open land only used in summer', from *sumor feld* it is first documented in 1218. Southenay is an Old English place name, *suthan ea* telling of its location as 'south of the river'. Stone Hill gets its name from one part of the hill, this is actually 'the stony slope' from *stan hielde*.

Selling

Domesday's Setlinges is the earliest record we have available. However the similarlty between that and the old record of the previous entry is clear. Indeed the meaning is the same 'those who shared a building' from Old English *sedl ingas* or 'the place at the building' from *sedl ing*.

Cheese Wood comes from the family of Richard Chese, here by 1327. Derived from Old English *here feld*, Harefield Farm refers to this as 'the place of the battle at the open land'. Although there is no record of any battle here, which may point to disputed territory rather than armed conflict. Owens Court has nothing to do with a personal name, here a corruption of *ofen* tells us there was once a furnace here.

Priviss Wood began as Privet's, the name derived from the family of John Priuet who were here by 1313. Listed as simply Rode in 1247 and coming from *rod* or 'clearing', Rhodecommon Wood has acquired references to the common land which was then given to woodland. Shepherd's Barn takes the name of the family of Jordain le Shepherde, recorded here in a document dated 1313,

The Sondes Arms public arms takes the name of the earls Sondes, a title created in 1880 and whose family seat was near here.

Sevington

Found in the Domesday record of 1086 as Seievetone, this comes from Old English *tun* with a Saxon or Jute personal name and describes 'the farmstead of a woman called Saegifu'.

Left: Fingerpost at Sevington. Right: Cheeseman's Green Lane at Sevington

St Mary's Church at Sevington

Shadoxhurst

Recorded as Schettokersherst in a document dating from 1239. Here the suffix represents Old English *hyrst* or 'wooded hill', unfortunately the first element has never been understood although most suggestions point to a personal or family name.

Blindgrooms Lane can be traced back to 1500 when it is seen as Blyndegromys, the name is from Middle English *blinde grome* 'the blind servant'. It seems unlikely this was an actual servant nor even a blind man. This probably represents a landowner, a nickname for someone who turned a blind eye to authority, blatantly ignoring rules and regulations.

Colebran Wood was home to the family Hamo Colebraund by 1315, named from the family. The lost place name of Minchin Court was once found here, a name from *myncen* or 'of the nuns', it was a possession of St James Hospital in Thanington. While the place name is now lost, the manorial reference is seen in Courthope Wood, the latter from Old English *hop* 'enclosed marshland'.

In the Kings Head we have a popular pub name. While a number of the signs depict a particular monarch, the origin of the name is a general show of support for the monarchy and showing a patriotic establishment.

Sheerness

A name from Old English *scir naess* and describing 'the bright headland', this is recorded as Scerhnesse in 1203. Although some sources suggest the alternative *scear* for a 'plough share' and alluding to the shape of the headland.

Ripney Hill is seen as Ripenay in a document dated 1192, this coming from a Saxon personal name and Old English *eg* and telling us it was once 'the dry ground in a marsh of a man called Rippa'. Trouts is a minor name which can be traced to 1334 when the family of Thomas Troutes was linked to this place.

Here the Jolly Sailor is a predictable name for a pub so close to the sea. Thus this makes for a very common name and the addition of an image of a happy chap not only gives the place a distinctive element but also suggests a good time can be had within. An invitation to the mariners who are not working is offered by the Ship on Shore. Another vessel is commemorated in the Victory Inn, probably the most famous in history after the Titanic, it was named to honour the men and officers who served aboard her as Nelson's flagship at the Battle of Trafalgar. Nelson is again the link to the Lady Hamilton, his famous lover and reputed to have been one of the most beautiful women ever to have lived. Yet another pub, the Nore, is linked to the ships around this coastline, this time named for the sandbank off Sheerness used as anchorage by the Royal Navy for almost two centuries.

Sheldwich

Found in 784 as Scilduuic, this comes from Old English *sceld wic* and describes 'the dwelling or farm with a shelter'.

Here we find Akhurst Farm, from *ac ersc* this describes 'the stubble field marked by an oak tree'. Cobrahamsole Farm has been recorded since the thirteenth century, this name speaks of 'the muddy pool of a man called Sceobba'. Copton is recorded as Coppanstan in 821 and as the modern form for the first time in 1208, this describes 'the *stan* or stone as the boundary marker of land of a man called Coppa'.

Halke House was built at a place called Halk in 1278, this represents Old English *halc* describing 'the small hollow'. Lees Court is the modern form of what began as *laes* or 'meadow'. Shepherds Hill comes from *sceap hyrst* or 'the wooded hill where sheep are grazed'. Stocking Wood is easy to see as coming from *stocket*, a Middle English term meaning 'a collection of tree stumps'.

Sholden

Seen in 1176 as Shoueldune, this comes from Old English *scofl dun* and refers to 'the shovel-shaped hill'.

Cottingtoncourt Farm began life as 'the farmstead of the cottagers', the later building of a much grander establishment saw the addition of 'court'. Foulmead tells us it is 'the meadow frequented by birds' from *fugol maed* and recorded from the thireenth century.

The Sportsman is not as common a pub name today. Once this advertised how these premises were where a number of sporting opportunities could be found, sometimes indoors and others representing the pub at football, cricket, angling, bowls, etc.

Sissinghurst

Found as Saxingherste towards the end of the twelfth century, this comes from a Saxon or Jute personal name and Old English *inga hyrst* and describes 'the wooded hill associated with the family or followers of a man called Seaxa'.

The Bell and Jorrocks public house at Frittenden

Sittingbourne

A name from Old English *side inga burna* which refers to 'the stream of the dwellers on the slope', this is recorded as Sidingeburn in 1200.

Bax is a name from Old English *byxe*, the earliest record is as Byx in 1327, simply referring to 'box trees'. Bayford Court took the name of 'the ford of a man called Babba', first seen as Babbeford in 1292. Blacketts is first recorded in 1782 and, while there is no record of such, can only have come from a family name. Chalkwell describes 'the spring from the chalky soil', from Old English *cealc wielle* and recorded since the fourteenth century.

Cheke's Court takes the name of the family of William Cheke, here by 1270. Nearby Tonge is recorded as Tangas in Domesday, Tanga in 1160, and Tange in 1190, this is from Old English *tang* 'the tongue of land'. To the east is Chestnut Street, which does not feature the Old English *straet* meaning 'Roman road' but would undoubtedly have been confused with this suffix considering the Roman influence in the county. Here the name is from Old French *chastaigniere* and recorded as Castayner in 1214, this describing 'the place where chestnut trees grow'. As late as the fifteenth century this was a thickly wooded area.

Cromer's Wood was known as such by 1598 when the family of James and William Cromer were living at Sittingbourne. From Old English *denu weg* and recorded as Denewei in the early thirteenth century, 'the way through a valley' is today known as Danaway. Fulston Manor is found as Fugeleston in 1197 and as Fulgeston in 1198, a name describing 'the *tun* or farmstead of a man called Fugul'. Woodstock is seen as Wodestok in 1202, this being from Old English *wudu stocc* and telling of 'the wood marked by stocks or stumps'.

Harman's Corner takes the name of William Hereman's family who were certainly here by 1235. Hempstead Lane comes from *henep stede*, Old English for 'the place where hemp is grown'. Highsted Forstal is derived from Old English *head stede* with the dialect *forstal* to speak of 'the high place in the front of farm buildings'. Howt Green is a corruption of *holt* or 'thicket', with *grene* a self-explanatory Middle English addition.

There is a modern tendency for place names to be drawn from the same source. Walk the newer estates in any town and find themes of birds, trees, and flowers. Developers feel the name of Elm Croft or Marigold Avenue gives a rural feel for the place even before the potential buyer has even seen the property. With pub names this is also the case, albeit to a lesser degree. Thankfully names such as the Beauty of Bath still exist. This is a variety of apple, one which would grow well in the Garden of England.

Key Street effectively means 'the quay on Watling Street', the *straet* being 'a Roman road' and the quay a wharf on a small stream. Longridge means exactly what it says, recorded as early as the eighth century this does indeed describes this settlement 'at the long ridge'. Milton Regis is a common name and one with two origins. Here *middel tun* speaks of 'the middle farmstead', with the addition a reminder this was a royal manor as evidenced by the record from 1052 as Middel tun paes cynges.

Murston, found since the late twelfth century, comes from *mores tun* or 'the farmstead of the moorland'. Quinton Farm remembers a family here by the fourteenth century, their name given as Quyntyn, Queynte, Quinte, and Quinton depending upon who made the record over the next three centuries. As a place name it first appears as Qwyntanstreet, showing it stands on a *straet* or Roman road. Yaugher is listed as Eldegar in 1334 and as Ealdgare in 1343, this being 'the old wedge-shaped area of land'.

Scuttington Manor features a Saxon personal name with old English *inga tun* to describe 'the farmstead associated with a man called Scuna', the name first seen in the thirteenth century. South Green has been seen since 1435, originally the 'southern common'. Westlands has hardly changed since the record of Westlande in 1278, this Middle English *west lande* and referring to 'the farmland west of the parish'.

In the case of the pub named the Three Hats we have a modern sign showing an image with three hats with plumes, much as those always depicted on the heads of Royalists during the Civil War era. Previously the sign showed a shako, a navy hat, and a billycock, as worn by a solider, a sailor, and a businessman. Just what the message was is unclear. War veterans are also commemorated in the Ypres Tavern, named after the five battles fought during the First World War.

The Prince Alfred is named after the second son of Queen Victoria, he was Duke of Edinburgh and pre-deceased his mother at the age of 56. Any pub named the Halfway House should not be taken literally, the message is not an indication of distance but an invitation to rest. Note in the USA the term refers to accommodation offered to prisoners as they are reintegrated back into society.

Small Hythe

Listed as Smalide and Smallhede in the thirteenth century, this name comes from Old English *smael hyth* and refers to 'the narrow landing place'.

Smarden

Found at the end of the twelfth century as Smeredaenne, this name comes from Old English *smeoru denn* and describes 'the woodland pasture where butter is produced'.

Left: Chequers public house at Smarden
Right: Smarden's village sign erected for the millennium celebrations

Fuller Farm is associated with the family of Thomas Fuller de Smarden, here in the middle of the fifteenth century.

Homersham Farm is derived from Old English *hamm* and a Saxon personal name and tells of 'the hemmed-in land of a man called Hunmaer'. Mayney Wood is easy to see as being associated with Richard de Maynye, here by 1348. Obeden can be traced to 1450 when the family of William Habynden were resident here. Stace Wood takes the name of former resident Thomas Stace, here before 1292. Westpherhawk Farm appears as Hwithsparroch in 1225 and as Whitesparok in 1379, this speaking of 'the enclosure of a man called Hwit'.

Omenden is recorded as Humindenn in 1254, a name from Old English *denn* with a personal name and telling of 'the woodland pasture of a man called Umma'. Romden is seen as Rumden in 1240 and as Romdenn in 1346, this being from *rum denn* and giving 'the wide or roomy woodland pasture'. Swift's Green has been known as such since the family of William Swyft were here in 1356. In Westhoy we have a name from *west haeg* and referring to 'the western hedged enclosure'.

Haffenden Quarter fingerpost near Smarden

Smeeth

A name meaning 'a smithy' from Old English *smiththe*, this is recorded as Smitha in 1018.

Locally we find Evegate, from Old English *theofa geat* or 'the way of the thieves'. It seems unlikely this refers to actual thieves, more likely a comment what was seen as unfair, perhaps a toll or land border.

Drinkers at the Woolpack public house may not be aware this was the name given to the large bale of wool weighing 240 pounds and which would have been a common sight on the roads as trains of pack animals travelled from place to place, with inns such as this acting as staging posts.

Snargate

Listed as Snergathe at the end of the twelfth century, this is derived from Old English *sneare geat* which describes 'the gate or gap where snares for animals are placed'.

Snave

An unusual name from Old English *snafa* which refers to 'the spits or strips of land', and is first recorded in 1182 as Snaues.

Stalisfield Green

Recorded as Stanefelle in Domesday and as Stealesfelde at the end of the eleventh century, this is probably from Old English *steall feld* and describes 'the open land with a stall or stable'. The addition is from Middle English *grene* and describes the village green.

Locally we find Holbeam, a name coming from *hol beam* and referring to 'the hollow where hornbeam grows'. Kennelling Farm can be traced back to a document dated 820 where it appears as Cyneuuoldincge and describes 'the place of a man called Cyneweald'. Redborough is either a manorial name, or one which comes from *read beorg* and tells of 'the red hill or mound'. Records are too late to show which is more likely.

Rushmere is listed as Ryssme in 1258, this from *risc mere* or 'the pool where rushes grow'. Windinghill Wood comes from Old English *wind hyll* which describes 'the windy hill'.

Stanford

A common name found in a number of places in England, with this example found in the modern form as early as 1035. This name comes from Old English *stan ford* and describes 'the stony ford'.

Butcher Wood is a reminder of 1357 when John and Walter Bochard were resident here. Hayton Manor was Haegtune in 1035, this from *gehaeg tun* 'the farmstead at the hedged enclosure'.

Staple

Listed as Stapel in 1240, this is derived from Old English *stapol* and describes the '(place at) the pillar of wood or stone'.

Barnsole is derived from *bern sol* 'the muddy pool near a barn'. Crixhall Farm comes from Middle English *crikeles hale* or 'the corner of land infested by crickets'. Summer Field comes from Old English *sumor feld* and telling of 'the open land only used in summer'.

The Three Tuns is found in the coat of arms of both the Worshipful Company of Vintners and the Worshipful Company of Brewers. A tun was a large cask used for distribution, containing some 250 gallons of liquid.

Stelling

A name which is probably from a Saxon or Jute personal name and Old English *ingas*. Recorded as Stellinges in Domesday, this appears to speak of the '(place of) the family or followers of a man called Stealla'. Nearby Stelling Minnis has an addition is from *maennes*, an Old English word meaning 'common land'.

Stelling Minnis sign

Boormanhatch Farm was worked by the family named Brouman, later seen as Borman in 1348, this being derived from *burmann* Old English for 'cottager'. Bower Wood comes from *bur*, Old English for 'a dwelling'. Lymbridge Green is found as Lemering in 1254, Limering in 1257, and Lymeryng in 1328. All these show the origin to be 'the place associated with a man called Leofmaer'. Mead Farm carries no surprises, this coming from Old English *maed* and tells us this began as a 'meadow'.

Souledge Farm is from Old English *suth dic* or 'the southern ditch', the earliest records dating from the fourteenth century. Whatsole Street took the name of the existing minor place name, this from *sol* with a Jute or Saxon personal name and speaking of 'the muddy pool of a man called Wada'. Wheeler's Street can be traced to the early seventeenth century, when the place was associated with the family of Elizabeth Wyllard. Whiteacre is derived from Old English *hwaet aecer* and speaks of 'the land on which wheat is cultivated'.

Stockbury

Domesday records this name as Stochingeberge in 1086, a name from Old English *stoc inga haer* and describing 'the woodland pasture of the dwellers at the outlying farmstead'.

Amets Hill is a reminder of the family who lived here by 1327, the document recording one William Amite. Pett is found as Pytte in the eleventh century and Pette by 1325. These show the origin to be Old English *pytt* 'pit or hollow'. Steels Wood can be traced back to 1296 when one Maurice Steghele was in residence, his surname speaking of 'the dweller by the stile'. West Wood speaks for itself as 'the wood to the west of the parish' and is recorded from the sixteenth century. The family known as De Wetesker were here in 1270, they are remembered by the present-day name of Whittaker's Wood.

Stodmarsh

The earliest surviving record of this name dates from 675, where it appears as Stodmerch. Here the name describes 'the marsh where the herd of horses are kept' and is derived from Old English *stod mersc*.

Locally we find Puckston Farm, from *puceles tun* or 'the farmstead of the goblins', clearly the place was considered haunted.

Stone

From Old English *stan* and meaning '(place at) the stones', this is recorded as Stane in the tenth century and as Estanes in Domesday.

Littlebrook Farm is not quite self-explanatory, this comes from *aet thaem lytlan broce* and describes its position 'at the little marshland'. Lord's Wood is named after former landholder William Lord, here around the end of the fourteenth century. Oxney is from *oxena gehaeg*, the appearing exactly as that in a document dated 1038, and describes 'the enclosure for oxen'. Stone Castle takes the name of Alan de Castello, here before 1278. Whole Farm is recorded as Hole in 1254, this being from *hol* and telling us this place lies in a 'hollow'. Swansfield Lodge was built on land originally associated with John Swon.

At the Ferry Inn we have a pub which not only advertises the crossing but offers hospitality for those waiting or having just landed.

Stour (River)

A river also known as the Great Stour, sharing a basic name also discussed under the Little Stour. As with its namesake this is recorded as Stur in 686 and Sture in 851. This is derived from a Celtic or British name referring to 'the strong one'.

Stourmouth

The earliest record of this name comes from the late eleventh century, where it appears as Sturmutha. This is derived from a Celtic or Old English river name and Old English *mutha* and refers to 'the mouth of the River Stour'. This river name probably means 'the strong one'.

Locally we find Pluck's Gutter, a name held to be a tiny part of the River Stour named after one Mr Pluck, a former landlord of the Dog and Duck Inn, who ferried workmen across the river during his time here and those men named it in appreciation of his efforts. The pub got its name from a sport which became very popular during the reign of Charles II. On the village pond the ducks had their wings pinned and thus, when the dogs were released, their only escape was by diving. On the subject of public houses we find the Rising Sun, a name of heraldic origins and chosen to represent many landed families and royals in the shape of Edward III and Richard III. It is one of the most common pub names in Kent.

Stowting

Records of this name include Stuting in 1044 and as Stotinges in the Domesday record of 1086. This is thought to be from Old English *stut ing* and describes 'the place noted for its lumpy hillock'. Alternatively the first element may be a Saxon or Jute personal name, in which case this would be 'the place associated with a man called Stut'.

The Black Horse is one of the most common of pub names. It is also one of the most popular devices in a coat of arms. The image has been used to represent many bodies and families and it is impossible to know which suggested itself as the name for a specific public house.

St Mary's Church at Stowting

Sturry

Documented as Sturgeh in 678 and as Esturai in the Domesday record of 1086, This name comes from Old English *ge* and a Celtic or Old English river name and tells us it was 'the district by the River Stour'. Again this river name is probably describing 'the strong one'.

Belce Wood is either Old English *bylce wudu* or Middle English *bilche wode*, although for once the language hardly matters for the meaning would be the same in 'the rounded wood', it is near a rounded hill. Blaxland Farm once belonged to the Blak or Blake family, hence the records of Blakeslaunde in 1226, Blakeslond in 1292, and Blakislonde in 1346. Breadland Farms is from *brad land* 'the broad cultivated land', found as Bradeland in 1226. Similarly 'the broad oak' from *brad ac* is now seen as Broadoak.

Hawe began as 'a hedged enclosure', one recorded from the late thirteenth century and derived from Old English *haga*. Heel Lane takes the name of the place recorded as Helde in 1327, *hielde* describing this as on 'a slope'. Hersden has been seen since the late thirteenth century, a name which began as speaking of the '(place of) the people of a man called Hersa'. Upstreet is first seen in 1690, a reference to the narrow road running up the hillside.

Island Road was described as Fresshe Eyland in 1502, that is 'the dry land surrounded by fresh water'. Kemberland Wood is the modern version of what began as 'the agricultural land of a man called Cynebeorht'. Mayton Farm, seen as Meytone in the thirteenth century, comes from *maegtha tun* and describes 'the farmstead where mayweed grows'.

Pub names include the Welsh Harp, the image of the harp is a link to the principality and a pointer to these premises having a connection to Wales, probably through a former landlord or owner.

Swale (River)

This river name is thought to come from Old English *swealwe* meaning 'rushing water'. The Swale was once an extensive river before the natural changes to the east coast.

Fowley Island is found in the estuary of the river, the name recorded since the thirteenth century and speaking of 'the island frequented by birds'.

Swalecliffe

Found in 949 as Swalewancliffe and in 1086 as Soaneclive, this comes from Old English *swealwe clif* and refers to 'the swallow's river bank', or perhaps 'the bank by the River Swale'.

Locally we find Chestfield Farm, a name from Old French *chestel vieil*, this being recorded as Cestevile in 1242 and Chesteuille in 1332. Probably a manorial name with the thirteenth century Cestevile family heralding from France.

Swingfield Minnis

The earliest surviving record dates from the late eleventh century as Suinafeld. This name comes from Old English *swin feld* and describes 'the open land where pigs are kept'. The addition here is Old English *maennes* or 'common land'.

The same suffix is seen in the records of Densole Farm, referred to in a document of 1539 as Densall Minnis and also Densall Bushes. These additions are references to 'common land' and 'shrubland' respectively. Densole comes from Old English *denn sol*, literally 'woodland pasture pool' although likely little more than a muddy pool at the end of the pasture nearest woodland. Smersole features the same suffix, here *smeoru sol* tells of 'the miry pool in rich pasture'. Foxholt has changed little since its earliest known record of Foxole in 1278. Indeed it has changed little from the original Old English *fox hol* or 'the hollow frequented by foxes'.

Boyington Court is found as Bointon in 1207, here 'the farmstead of the young men' comes from Middle English *boiena tone*. The favourite kind of name for the author is that which provides a shapshot of a moment in time, an image of Saxon life which no painter would paint and no camera could capture. Brandred Farm is one such place name, albeit one which would be unlikely to be considered picturesque. Old English *brand ried*, appearing as simply Brand in Domesday, describes 'the land cleared of undergrowth by burning', instant imagery.

Hoad Wood grows alongside the *haeth* or 'heathland'. Ridge Row does indeed refer to 'a ridge of land' here, the name from *hrycg* and seen as Regge in 1226.

CHAPTER NINETEEN – T

Tankerton

Found as Tangerton in 1240, this name comes from an Old German personal name and Old English *tun* and refers to 'the farmstead of a man called Tancred or Thancrad'.

Tenterden

Recorded as Tentwardene in 1179, this name is derived from Old English *ware denn* with the place name Thanet, a name discussed under its own entry. This name refers to 'the woodland pasture of the dwellers at Thanet'.

Ashbourne Mill Road is a reminder of the 'stream among the ash trees' which was later utilised by a watermill. Ashenden describes 'the *denn* or woodland pasture of a man called Aeschere'. Belgar is from Old French *bel gard*, which can still be seen as 'beautiful garden'. Bugglesden is from *boc holtes denn* 'the woodland pasture by the thicket of beech trees'. Burgess Wood reminds us of the Burgeys family, here in 1292. Collington Wood was home to Nathaniel Collington, vicar of this parish 1662–82.

Craythorne is found as Crawethorn in 1226, this coming from Old English *crawe thorn* 'the thornbush frequented by crows'. Dawbourne Wood comes from Middle English *dufe burne* 'the stream where doves are seen'. From an earlier name for a *burna* or 'stream' comes the name of Dumbourne, recorded since the thirteenth century as a place name, it describes 'the stream of the Duningas' a tribal name describing 'the dwellers on a hill'. A Saxon personal name and Old English *burne* comes Huntbourne, a name meaning 'the stream of a man called Hunta'.

Huson Farm is found as Husneah in 863, this telling of 'the river of a man called Husan' where the personal name is suffixed by *ea*. Ingleden is found from the thirteenth century, this being the *denn* or 'woodland pasture of a man called Igilwulf'. Kench Hill is not seen before the sixteenth century although if we accept the name is much older this refers to 'Kent's hill', this being next to the traditional border with Sussex. Wellington Place has clearly been influenced by one of our nation's most important figures, the Duke of Wellington. However the name predates him by at least six centuries, for this was recorded as Wyludenn in 1292 and represents 'the woodland pasture of a man called Wilfrid'.

Morghew is an unusual sounding place name but one with the wondrous meaning of 'morning gift' from Old English *morgen giefu* and referring to a piece of land officially given by a husband to his bride on the morning after they married. Pick Hill is seen as Pikhilde in 1491, from *pic hielde* this is 'the slope running to a point'. From Middle English *shrubbe cot* and meaning 'the cottage near the shrubs' comes the name appearing on modern maps as Scrubcut Wood.

In the Vine Tavern we find one of the oldest English pub names known. Brought to our shores by the Romans it of course refers to the fruit used to produce wines. Once known as the Black Horse, the William Caxton was renamed in 1951 at the Festival of Britain. Caxton is remembered as the man who, in 1475, printed the first book in English. However the sign shows an image of the next book, entitled *The Game and Playe of Chesse*.

Teynham

A name appearing as Teneham in 798 and as Therham in 1086, this comes from a Saxon or Jute personal name and Old English *ham* and describes 'the homestead of a man called Tena'.

Barrow Green comes from *beorg* the 'mound or barrow', it is recorded as simply Bergh in 1327. Conyer is from an Old French term, *conniniere* describing 'the rabbit warren'. Deerton Street is also a reference to food animals, from *deor tun* 'the farmstead where wild animals are caught' and recorded as Dertone in 1334. Frognal is found as Froggenhale in 1279, from Old English *froggena halh* 'the corner of land where frogs are found'. Tickham Lane is recorded as the minor place of Tykehamme in 1327, from *ticcena hamm* here is 'the hemmed-in place where young goats are kept'.

Thames, River

An ancient river name and one where the earliest record of Tamesis from 51bc is one of the oldest known for any English place name. Indeed, so old is this name it is not clear if the meaning is 'the dark one' or simply 'river', although both allude to moving water.

Thanet

A name found as early as the third century, when it appears as Tanatus, with the later Domesday record of Tanet. This is undoubtedly a Celtic name, one which most likely means 'bright island' and thus referring to a beacon'.

Thanington

Recorded as Taningtune in 833, this name is probably telling of 'the farmstead associated with a man called Tan', with the Saxon or Jute personal name followed by Old English *ing tun*.

Cockering Farm is derived from Old English *cocer ingas* meaning 'the people of the quiver or sheath', a clear reference to a Saxon or Jute tribe known for carrying bows and arrows. Hand Wood is a corruption of Honywood, this the family who once owned neighbouring land. Iffin Farm is recorded as Yethyngge and Ithynge in the fourteenth century, from Old English *geoguth ingas* this is the '(place of) the youthful dwellers'.

Tonford Farm is derived from Old English *between fordum* which is easier to see as this place being 'between the fords'.

Throwley

Domesday lists this name as Trevelai in 1086, a name which is derived from Old English *thruh leah* and describes 'the woodland clearing by a conduit'.

Back Shaw is of manorial origin, a fourteenth century family recorded as both Bak and Bac. Bells Forstall, found in 1782 as Bellhorn, takes the name of the de Belle family who were here by 1327 and adds the dialect word *forstal* referring to land in front of farm buildings. Gardener's Wood took the name of the family of John and Roger le Gardiner, recorded as living at Throwley in a document dated 1270.

Hayward's Hill is associated with the family of John le Heyward in 1292. Wilgate Green is found as Wilgare in 1235, as Wylgar in 1270, and as Wyligare in 1327. These are derived from Old English *wilig gara* and telling of 'the triangular plot on which willows grow'.

Tilmanstone

Found in the Domesday record as Tilemanestone, this name features a Saxon or Jute personal name and Old English *tun* and describes 'the farmstead of a man called Tilmann'.

Barville Farm is found as Barefeld in 1313 and as Barfelde in 1346, this is from *baer feld* or 'the open land near a pasture for swine'. Lower Longlands is from old English *lang fryhthe* and describes 'the long woodland'. Thornton Farm is seen as Thornton as early as 1240, this coming from Old English *thorn tun* and describing 'the farmstead marked by thorn bushes'.

The Plough and Harrow is almost as rural as a pub name can be. With the plough used to cut through and turn the soil, the harrow was a framework dragged across the same area to break up the soil which would reduce the clumps to a finer tilth and give the seeds a better chance of germinating. As a pub name it offers a welcome to those who worked the land which, at the time it was coined, was almost everyone.

Tunstall

A name found in a number of counties of England. This is recorded as Tunestelle in Domesday and is derived from Old English *tun stall* and describes 'the place of the farmstead'.

Bex Wood is a lasting reminder of a manorial name from the fourteenth century, the family recorded as both Bak and Bac. Their family name comes from Old English *baec* and describes 'the dwellers on the ridge'. Ufton Court is found from the thirteenth century, the suffix *tun* following a Saxon personal name and giving 'the farmstead of a man called Uffa'.

Tyler Hill

The earliest surviving record dates from 1304 as Tylerhelde. This is from Old English *tiglere helde* and describes 'the slope of the tile makers'.

CHAPTER TWENTY – U

Upchurch

Documented as Upcrycean at the end of the twelfth century, this name comes from old English *upp cirice* and refers to 'the elevated church'.

Bayford is first seen in 1442 as Boywurthe, this represents Middle English *boian worthe* or 'the enclosure of the young men'. Horsham is found exactly as the modern form as early as 1382, this from Old English *hors hamm* and speaking of 'the hemmed-in land where horses are grazed'. Wetham Green is derived from *waet hamm* 'the wet water meadow' and recorded as Weteham in the fourteenth century. Otterham Creek is a place name and one which has taken the name of the tidal creek on which it stands. Here, from Old English *attor* is the water course describes as 'the swollen one'. Nearby is the Twinney Creek, an inlet from the sea named for it being 'between the rivers'.

Upper Hardres

A name found as Haredum in 785 and as Hardes in 1086. Here is a name derived from Old English *harad* and describes the '(place at) the woods'. There is also a Lower Hardres.

Fingerpost at Lower Hardres

The Granville public house at Lower Hardres

The evolution of Bossingham began with Bossingcamp in 1226, Bosingkomp in 1264, and Bosyncompe in 1343. Pronunciation of -camp later became -cum, then -um, and -em, and finally -ham. Hence the suffix is not *ham* but *camp* and refers to 'the enclosed land of a man called Bosing'.

Broxhall Farm is seen as Brockyshole in 1304 and as Brokkeshole in 1338, names from Old English *brocces hol* 'the hollow frequented by badgers'. Lynsore comes from Old English *hlinces ora* and, recorded as Linchesora in 845, speaks of 'the bank on the rising ground'. Stockfield Wood takes the name of Henry de Stoke, and adding *feld* or 'open land'. Street End speaks for itself as 'the termination of the Roman road'.

CHAPTER TWENTY ONE – W

Wantsum (River)

Originally the channel which splits the mainland from the Isle of Thanet, most likely this describes the erratic nature of its course in a name meaning 'to wend'.

Warden

Listed as Wardon in 1207, this name comes from Old English *weard dun* and describes 'the lookout hill'.

Warehorne

Found as Werahorna in 830 and as Werahorne in 1086, this name speaks of 'the horn shaped piece of land by the weirs' and is from Old English *wer horna*.

Capel Farm is recorded as Capella in 1275 and Chapele in 1292, this is from Middle English *capel* or 'chapel'. Faggs Wood was associated with the Fagg family, at neighbouring Orlestone by 1613 and recorded at Warehorne two centuries later. Longrope Wood does indeed refer to a 'long rope' but should be seen as a measurement of the land although the significance of such is not fully understood. Orlestone is found in Domesday as Orlavestone, this coming from a Saxon personal name and Old English *tun* to give 'the farmstead of a man called Ordlaf'. Seen as Titentone in the Domesday record of 1086, Tinton began as 'the farmstead of a man called Tida'.

Outside the World's Wonder the sign painter has used his imagination to depict a cockerel having laid a square egg, which is hatching to reveal a bottle of beer within. The real origin is probably an even better story. In the middle of the nineteenth century two empty and very run down cottages were bought by local man Tom Night. Not known as a man of great means everyone wondered where he had got the money and before long, with Tom working every spare minute on his investment, he was the talk of the village. At first Tom ignored the chatter but eventually, tired of the incessant gossip, turned on his neighbours and yelled: "You wonder where I got the money? You all wonder what is being built? Everyone wonders how I got the licence? Now the world wonders what Tom is going to call it? Well the world can stop wondering for there can be no more apt name than the World's Wonder!"

Watling Street

One of the four major roads covered by William the Conqueror's laws promising safe passage for travellers. This very political gesture was the reason the name has been transferred by others throughout England, although there is no physical connection between them. This is a Roman road, roughly following the line of the modern A5 trunk road between Dover and Holyhead off Anglesey, although the Roman part only reached Wroxeter.

Listed as Waetlinga straet in 880, Waeclinga straet in 926, and Watlingan straet in 957, this name can be traced back to one very small stretch of this 240 mile road. That small part was around the place name recorded as Waecelingaceaster around the end of the ninth century, a name meaning 'the Roman station of the people of Wacol', an old name for what is now known as St Albans.

Westbere

A name meaning 'the westerly woodland pasture' from Old English *west baer* is recorded as Westbere in 1212.

Westgate on Sea

Listed as Westgata in 1168, this name comes from Old English *west geat* and refers to 'the westerly gate or gap'.

The obviously French Dent-de-Lion is from the family of William de Dandelyon, here by 1382. Originally a French place name, literally 'lion's tooth', is pronounced 'dandelion' and is the origin of this ubiquitous weed's name. Fleet is from *fleot*, a name describing 'a tidal creek'.

Westwell

From Old English *west wella* and describing 'the westerly spring or stream', this name is seen in 1086 as Welle and as Westwell in 1226.

Digge Farm and Dignash are local names showing an association with the Digge family from the reign of Edward III. The latter suffix tells of a prominent ash tree marking this place. A family recorded as Elfegh in the fourteenth century will have given their name to the minor place name of Elvy. Longbeech Wood is first recorded in the thirteenth century, a name from *lang bece* describing 'the long woodland of beech trees'. Witchling Wood gets its name from Old English *wice leah* and tells of 'the woodland clearing marked by wych elms' and recorded as Wicheleg and Wichelegh in the early part of the thirteenth century.

Whitfield

A name derived from *hwit feld* and referring to 'the white open land'. This name appears in a document dated 1228 as Whytefeld.

The Drove is a modern name with an ancient name, coming from Old English *draf*, one referring to this as part of the drover's route and used much earlier than the first record of 1292. Kearnsey is one of those rare place names from Old French, here *cressonniere* describing 'the place where cress grows'. Lenacre is from *lin aecer* or 'the cultivated land where flax is grown'.

A name recorded as Napisherst in 1198, Napesherst in 1275, and Napserst in 1347 shows this is derived from Old English for 'the *hyrst* or wooded hill of a man called Naep'. However since the fifteenth century the name abruptly became Napchester, suggesting an alternative suffix of *caester* and giving 'the Roman stronghold of a man called Naep'. It has been suggested this abrupt change from –hurst to –chester was deliberate, especially considering the *caester* suffix is Old English and therefore pre-Norman and yet appears to have remained hidden until 1451. Unless the present suffix was hidden for seven or eight centuries and records of the earlier suffix have simply failed to survive, the argument of a deliberate change does hold water. Yet it does not explain why and this, as when defining all place names, is the more important factor.

Pineham is found in the Domesday record of 1086, this coming from Old English *pinn ham*. The literal meaning here is 'peg or pin settlement', which seems to be telling us the enclosure in which the place was found was marked by pegs of some description. A tribal nickname meaning 'the lazy ones', probably meaning they were known for their tardiness, was recorded as Saenling in the eighth century and as Singledge today.

The local pub is the Archer. Today the pub sign shows the quite ludicrous image of a knight in armour aiming a longbow. Previously the bowman is depicted as the more plausible huntsman. Yet neither of the images are correct for this is an old manorial name. As with nearby Archer's Court this is named after the family who held this manor from the thirteenth century, their name recorded as le Archier, Archer, and Archyer.

Whitstable

Both eleventh century records of Witestaple and Witenestaple point to an origin of *hwit stapol*, giving a meaning of 'the white posts'. Clearly such were used to mark something, probably a safe route.

Balserstreet Farm takes its name from a local family recorded here from the thirteenth to the fifteenth centuries. The name is uncertain for it is found as Belsyre, Belsire, Balsier and Balser, while the addition of 'Street' came some time later. Benacre, found as early as the ninth century, comes from Old English *bean acer* 'the cultivated land where beans are grown'.

Bigbury Camp is said to be where the Romans under Julius Caesar engaged the Belgae tribe in 54bc, this was not the Roman Conquest of Britain which did not begin until ad43. When this battle was fought the place was certainly not called Bigbury, for this was not coined until the arrival of the Saxons, Angles and, in Kent, the Jutes. We have no idea what the fortification was called in those days, however the Old English name was *bycge burh* or 'the bulging stronghold', the bulge being the hill on which the fort was constructed.

Fox's Cross Hill and Fox's Cross Road take the name of John le Fox, whose family are known to have been here by 1278. Gorrell Bridge is a reminder of how polluted streams are not only a modern problem for the name tells us this was 'the dirty stream'. Recorded as Gorewelle in 1278, this is from *gor wielle* and, while it may describe discolouration of the water, it probably shows this was used as an open sewer since *gor* can also be used to mean 'dung, filth'. Lypcatt Wood gets its name from the Lipyeatt family, landholders in the area by the eighteenth century. Wraik Hill is recorded as Rake in the thirteenth century, this coming from *hraca* and describing a topographical feature said to be resembling 'a throat, gulley'.

Pub names begin with the Prince Albert who, during the nineteenth century, was at least as popular as his wife, Queen Victoria. The Fountain is used in heraldry to refer to the Plumbers' Company and also, as here, the Master Mariners. The Smack Inn refers to the coastal location, a 'smack' being the traditional fishing boat, while the Old Neptune takes the name of the Roman god of the sea. The Long Reach is a pub built in the 1960s to replace a pub abandoned ten years earlier as it was repeatedly flooded in its location on a long reach of the River Dart.

Wichling

Found as Winchelinge in 1220, this comes from a Saxon or Jute personal name and Old English *ingas* to tell of the '(place of) the family or followers of a man called Wincel'. Earlier this is seen as Winchelesmere in 1086, the Domesday record showing it was earlier referred to as 'Wincel's pool'.

Filmer Wood is listed as Ffulme in 1258 and as Filmer for the first time in 1498. Here Old English *ful mere* describes 'the foul or dirty pool'.

Wickhambreaux

Records of this name include as Wicham in 948, as Wicheham in 1086, and as Wykham Breuhuse in 1270. The name comes from Old English *wic ham* and describes 'the homestead with a vicus from an earlier Romano-British settlement'. The addition here is manorial, a reference to the de Brayhuse family, here by the thirteenth century.

Locally we find Frognall Farm, a name originally seen as both North Frogenhole and South Frogenhole in 1360, this comes from *froggena hol* and speaks of 'the hollow frequented by frogs'. Sandpit Cottages and Sandpit Lane share a name which still explains itself although coined by the late thirteenth century. Supperton Farm is seen as Suthburton in 1242, this from *suth bere tun* or 'southern farmstead where barley is grown'. Note 'barley' was also used as a generic term for any cereal crop.

Waterham Farm is derived from Old English *waeter hamm* and tells of 'the water meadow'.

Willesborough

The earliest surviving record of this name dates from 863 where it appears as Wifelesberg. Here a Saxon or Jute personal name precedes Old English *beorg* and refers to 'the hill or mound of a man called Wifel'.

Bockham takes the suffix *hamm*, here understood as 'the water meadow of a man called Bocca'. This element *hamm* has been seen for several kinds of natural feature and was once misleading. However when one similarity was spotted on high definition contour maps, it was realised these were describing 'hemmed-in land', approachable on fairly even ground from just one side.

Conningbrook comes from *cyne broc*, the 'royal marshy ground'.

No surprise to find so many public houses named after the traditional emblem of the county of Kent, the White Horse Inn here is just one. Similarly the Windmill is a predictable pub name, for the flat lands of the east of England proved the ideal location for wind-power to replace the earlier watermills after the twelfth century. William Harvey was born in Folkestone in 1578, he is best known for his discovery of the circulation of the blood (although his contemporaries mocked him for the claim) and as the personal physician of Charles I.

In the Churchill we find the name of the former prime minister who spoke the famous line "Never, in the field of human conflict, has so much been owed by so many to so few," when referring to the RAF fighters' role at the Battle of Britain fought in the skies over Kent. Both the man and the battle are depicted by the sign. One of the enduring songs of the Second World War is also linked to the Albion pub name. This was a traditional name for Britain, derived from *albus* the Latin for 'white' and a reference to the first sight to greet those from the Continent and which also produced that famous song *White Cliffs of Dover*.

Wingham

Listed as Uuigincggaham in 834 and as Wingheham in 1086, this comes from Old English *inga ham* with a Saxon or Jute personal name and describes 'the homestead of the family or followers of Wicga'. However there are some who give the first element as Old English *wig*, in which case this is understood as 'the homestead of the followers at the heathen temple'.

Blackney Hill appears to have taken the name of 'a dark island in the marshy ground' created by the regular flooding of the Wingham river. Bramling, which must always be pronounced without the final 'g', has been recorded since the twelfth century and comes from *bremelingas* 'the dwellers among the brambles'. Crockshard Farm has a name found since the thirteenth century, this being *ceom crocseard* or 'pot sherd', a piece of broken pottery. This name is not contemporary but would have been an early reference to where broken pottery had been unearthed and, in order for the name to stick, plenty of it which would indicate a potters from pre-Saxon or Jute times.

Dam Bridge is a comparatively late name, recorded as Danne Bridge in 1790. This probably represents Middle English *dene* or 'valley', the bridge crossing the stream in that valley. Nash comes from Old English *aet thaem aesce* and later Middle English *atten asshe* meaning 'at the ash tree', the migration of the final 'n' from the end of *atten* to *asshe* is not only common but should almost be expected.

Neavy Downs and also Neavy Cottage can probably be traced back to William le Neve, listed as a resident of Wingham in 1313. Great Rusham, Little Rusham and Rusham Farm all share a common origin in Old English *risc hamm* and 'the water meadow where rushes grow'. Shatterling is the only remaining clue as to an earlier name for what is now the River Wingham, with Middle English *schater* quite literally meaning 'shatter' and here used to describe a noisy stream.

Trapham Farm was recorded as Tropham in 1270 when that place was known for being 'the homestead of a man called Troppa'. Walmestone appears as Wielmestun in the eleventh century, this speaking of 'the farmstead of a man called Wighelm'. Both Great Wenderton Farm and Little Wenderton Farm share an origin in a place name recorded as Wenderton and Wendreton in the thirteenth century. Here Old English *tun* follows a Jute or Saxon place and speaks of 'the farmstead of a man called Wenthry'.

Wittersham

Listed as Wihtriceshamme in 1032, this name comes from a Saxon or Jute personal name and Old English *hamm* and refers to 'the hemmed in land of a man called Wihtric'.

Acton Farm comes from Old English *ac denn*, telling of 'the pasture by the oak trees'. Owley is derived from *ule halh*, Old English for 'the corner of land frequented by owls'. Blackbrook describes 'the brook flowing through the marshy land'. Palstre Court has changed little since the original English *palester* and refers to 'the point or spit (of land)'. Peening Quarter is found as

Le Pendynge in 1440, this from *pynding* or 'enclosed place'. The addition, first seen in 1790, means exactly what it says.

Womenswold

A name which shows the necessity for checking early forms for this has nothing female about the name. The earliest form dates from 824 when the name is given as Wimlincgawald, this shows the name to be from a Saxon or Jute personal name and Old English *inga weald* and telling us this was 'the forest of the family or followers of a man called Wimel'.

Locally we find Denne Hill, which has changed little since documented as Denne in 1226 and as Denhill in 1610. The basic name comes from Old English *denn* or 'woodland pasture'. A document dated 1347 shows a family by the name of De Ffyneaws living here, their surname is the only candidate for the origin of Finnis Wood. Nethersole Farm is a common enough English place name, from *nithera sol* it describes an even more common sight in 'the muddy pool'.

Woodchurch

A name from Old English *wudu cirice* and referring to 'the church by the wood'. This name is recorded as Wuducirce around the end of the eleventh century.

Berridge Farm is first seen in 1278 as Belgeregg, this being from 'the *hrycg* or ridge of land associated with a man called Belga'. Broadshaves is derived from the family of William Broadsheafe, first recorded here in 1591. Bold Snoad Wood is first seen in 1240 as Boghesnod, later as Bosnod in 1254 and as Bowsnode in 1313. Here the suffix *snad* tells us of 'the detached woodland of a man called Bogan'.

Cold Blow is a corruption of the name of the former hundred, this being the meeting place of the Cornilo Hundred. Its name comes from Old English *cweorn hlaw* 'the hill with a quern or handmill'. Hell Wood is recorded as Helde in 1261, showing this to be derived from *hielde* and referring to 'the slope'. Little surprise to find King Farm was from *cyne tun* 'the royal farmstead', although the farm is all that remains of the name or the place which once appeared on the map as Kennetune in Woodchurch.

Newhurst is not found before 1720, probably a rough indication of when the 'new wood' was planted. While records of the name begin in the eighteenth century, it seems there can be no other origin for Redbrook Street than *hreod broc straet* 'the marsh of reeds by a Roman road'. Robhurst is recorded as Bubehurst in 1254 and Bubherst in 1278, the name from 'the *hyrst* or wooded hill of a man called Bubba'.

Woodnesborough

Listed as Wanesberge in Domesday and as Wodnesbeorge at the end of the eleventh century, this is derived from Old English *beorg* and the name of a pagan god. The locals must have felt this a remarkable place for they knew it as 'the mound associated with the god Woden'.

Christian Court can be traced back to 1357 when the name is recorded as being associated with one Jocelyn St Cristian. Flemings is one of the many minor place names which began as a manorial name, the Fflemyng or Flemyng family were recorded here in the fourteenth century. Hammill is a contraction of Old English *hamm weald*, 'the high forest near the hemmed in land' being recorded as Hamolde in 1086. Ringlesmere Road is first seen in the record of Ryngwynme in the thirteenth century. Here Old English *mere* follows a Saxon personal name and tells of 'the pool of a man named Hringwynn'.

Pubs named the Poacher seem to point to a known criminal, and yet early names are rarely as obvious. Hence perhaps we are looking at a name which comes from the old adage "old poachers make the best gamekeepers" and suggests how the best landlords come from those who most enjoyed a drink.

Wormshill

As with the previous name, this is derived from a pagan god, here with Old English *hyll*, and speaking of 'the hill of the god Woden'. The name is recorded as Godeselle in the Domesday record of 1086, as Wotnesell around 1225, and as Worneshelle in 1254.

Here we find Savage Wood, certainly the name of a horror story of the future but royalties are due to the family of Arnold Salvage, here by the middle of the thirteenth century.

Worth

A name derived from Old English *worth*, an element usually seen as a suffix. The name describes 'an enclosure', a reference to a structure designed to keep the livestock and belongings in rather than a defensive construction. The earliest surviving record dates from 1226 as Wurth.

By far the most common colour seen in a public house name is 'blue'. Indeed there are so many, and so very few where any part of the building is actually blue, there must be a different meaning and research shows there to be two. Sometimes it is the colour associated with Christianity, but most often it shows a political allegiance, that to the Whigs forerunner of the Liberal Party. Hence the name of the Blue Pigeons.

Wye

A name which tells us this was a special place for the Germanic peoples who came to these shores in the fifth century ad. The name comes from Old English *wig* and describes the '(place at) the heathen temple' and is recorded as Uuiae in 839 and as simply Wi in Domesday.

Above: A traditional sign indicating this is Wye .....

..... and a more imaginative version

Amage Farm is derived from Old English *hamming* 'the place of the hemmed-in land'. Browning Bridge is a lasting reminder of the lost place name of Bronesford, itself meaning this was 'the river crossing of a man called Bran'. Fanscoombe Wood is listed in Domesday, albeit simply as Fanne, from *fenn* obviously 'fenland'. Not until the fourteenth century do we see the additional *combe* meaning 'valley', while the addition is very recent. Lavington Farm puts *denn* after a personal name and describing 'the woodland pasture of a man called Lafa'.

Marriage Farm is not what it seems, this comes from *maere hrycg* and actually refers to 'the boundary ridge'. Palmstead refers to 'the homestead near where plums are grown', from Old English *peru hamstede* and appearing as Perhamstede as early as 747. Plumpton Farm is seen as Plumton in 1272, a name from *plume tun* meaning 'the farmstead at the plum trees'. Sillibourne Farm is named after the land and the water course flowing through it in a name meaning 'the chalky stream'. Soakham Farm, first seen as Saecumb in 824, is from *sacu cumb* and tells of 'the disputed valley'. Listed as Wytherestun in 1272 and as Wythereston in 1307, in Withersdane we find a Saxon personal name and Old English *tun* telling us this was 'the farmstead of a man called Wither'.

In the New Flying Horse we have a pub name where the 'new' tells of a new build or refurbishment. The basic name is of course a reference to Pegasus of Greek mythology, this being adopted by the Knights Templar in their coat of arms and probably coming here from this.

CHAPTER TWENTY TWO

Element	Origin	Meaning
ac	Old English	oak tree
banke	Old Scandinavian	bank, hill slope
bearu	Old English	grove, wood
bekkr	Old Scandinavian	stream
berg	Old Scandinavian	hill
birce	Old English	birch tree
brad	Old English	broad
broc	Old English	brook, stream
brycg	Old English	bridge
burh	Old English	fortified place
burna	Old English	stream
by	Old Scandinavian	farmstead
ceap	Old English	market
ceaster	Old English	Roman stronghold
cirice	Old English	church
clif	Old English	cliff, slope
cocc	Old English	woodcock
cot	Old English	cottage
cumb	Old English	valley
cweorn	Old English	queen
cyning	Old English	king
dael	Old English	valley
dalr	Old Scandinavian	valley
denu	Old English	valley
draeg	Old English	portage
dun	Old English	hill
ea	Old English	river
east	Old English	east

ecg	Old English	edge
eg	Old English	island
eorl	Old English	nobleman
eowestre	Old English	fold for sheep
fald	Old English	animal enclosure
feld	Old English	open land
ford	Old English	river crossing
ful	Old English	foul, dirty
geard	Old English	yard
geat	Old English	gap, pass
haeg	Old English	enclosure
haeth	Old English	heath
haga	Old English	hedged enclosure
halh	Old English	nook of land
ham	Old English	homestead
hamm	Old English	river meadow
heah	Old English	high, chief
hlaw	Old English	tumulus, mound
hoh	Old English	hill spur
hop	Old English	enclosed valley
hrycg	Old English	ridge
hwaete	Old English	wheat
hwit	Old English	white
hyll	Old English	hill
lacu	Old English	stream, water course
lang	Old English	long
langr	Old Scandinavian	long
leah	Old English	woodland clearing
lytel	Old English	little
meos	Old English	moss
mere	Old English	lake
middel	Old English	middle
mor	Old English	moorland
myln	Old English	mill

niwe	Old English	new
north	Old English	north
ofer	Old English	bank, ridge
pol	Old English	pool, pond
preost	Old English	priest
ruh	Old English	rough
salh	Old English	willow
sceaga	Old English	small wood, copse
sceap	Old English	sheep
stan	Old English	stone, boundary stone
steinn	Old Scandinavian	stone, boundary stone
stapol	Old English	post, pillar
stoc	Old English	secondary or special settlement
stocc	Old English	stump, log
stow	Old English	assembly or holy place
straet	Old English	Roman road
suth	Old English	south
thorp	Old Scandinavian	outlying farmstead
treow	Old English	tree, post
tun	Old English	farmstead
wald	Old English	woodland, forest
wella	Old English	spring, stream
west	Old English	west
wic	Old English	specialised, usually dairy farm
withig	Old English	willow tree
worth	Old English	an enclosure
wudu	Old English	wood

CHAPTER TWENTY THREE

Oxford Dictionary of English Place Names by A.D. Mills
The Concise Oxford Dictionary of English Place-Names by Eilert Ekwall
A Dictionary of Pub Names by Leslie Dunkling and Gordon Wright
The Place Names of Kent by Judith Glover

BV - #0040 - 090322 - C0 - 234/156/11 - PB - 9781780912301 - Matt Lamination